BASICS OF CELL BIOLOGY

BASICS OF CELL BIOLOGY

SACHIN PADGAONKAR

ANMOL PUBLICATIONS PVT. LTD.
NEW DELHI - 110 002 (INDIA)

ANMOL PUBLICATIONS PVT. LTD.

H.O.: 4374/4B, Ansari Road, Darya Ganj,
New Delhi-110 002 (India)
Ph.: 23278000, 23261597

B.O.: No. 1015, Ist Main Road, BSK IIIrd Stage
IIIrd Phase, IIIrd Block
Bangalore - 560 085 (India)
Visit us at: www.anmolpublications.com

Basics of Cell Biology

ISBN 978-81-261-3494-6

PRINTED IN INDIA

Printed at Mehra Offset Press, Delhi.

Contents

Preface

Modern biology is rooted in an understanding of the molecules within cells and of the interactions between cells that allow construction of multicellular organisms. The more we learn about the structure, function, and development of different organisms, the more we recognize that all life processes exhibit remarkable similarities.

Basics of cell Biology concentrates on the macromolecules and reactions studied by biochemists, the processes described by cell biologists, and the gene control pathways identified by molecular biologists and geneticists. In this millennium, two gathering forces will reshape molecular cell biology: *genomics,* the complete DNA sequence of many organisms, and *proteomics,* a piece of knowledge of all the possible shapes and functions that proteins employ.

Preface

Chapter 1

The Dynamic Cell

To study the properties of the molecules of life and the innumerable variations on basic themes that are found in different organisms, modern researchers employ concepts and experimental techniques drawn from biochemistry, molecular biology, genetics, and cell biology. The resulting discipline of *molecular cell biology* investigates how cells develop, operate, communicate, and control their activities, and on occasion go awry. This book contains the authors' attempt to describe systematically the current state of knowledge about cells and to present many of the key experiments that have led to our current understanding of cellular life. It may seem to you, our new reader, a daunting challenge. Our hope is that the overwhelming inventiveness and sheer beauty of construction of biological systems will intrigue you and amply reward your efforts to understand the story we tell.

Living systems, including the human body, consist of such closely interrelated elements that no single element can be fully appreciated in isolation from the others. Organisms contain organs; organs are composed of tissues; tissues consist of cells; and cells are formed from molecules. The unity of living systems is coordinated by many levels of interrelationship: molecules carry messages from organ to organ and cell to cell; tissues are delineated and integrated with other tissues by noncellular membranes secreted by cells; and cells gain identity from contact with other cells. Generally all the levels into which we fragment biological systems interconnect. To learn about biological systems, however, we must take a segment at a time. The biology of cells is a logical starting point because an organism can be viewed as consisting of interacting cells, which are the closest thing to an autonomous biological unit that exists. The integration of cellular activity into tissues, the development of organisms by growth and specialization of cells, and the metabolic events fueling the dynamism of living systems are all topics on which we will touch, but they are all topics that fall within the province of other subdisciplines of biological science.

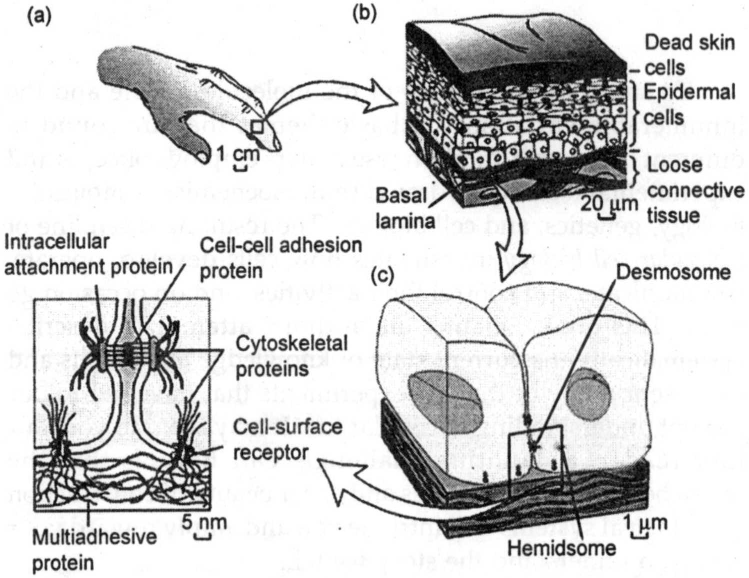

Fig. Living systems such as the human body consist of closely interrelated elements. (a) The surface of our hand is covered by a living organ, skin, that is covered by several layers of tissue. (b) An outer covering of hard, dead cells protects the body from injury, infection, and dehydration. This layer is constantly renewed by living epidermal cells, which also give rise to hair and fur. Deeper layers of muscle and connective tissue give skin its tone and firmness. (c) Tissues are formed through subcellular adhesion structures that join cells to each other and to an underlying layer of supporting fibers. (d) At the heart of the adhesion are its structural components: phospholipid molecules that make up the plasma membrane and protein molecules which form strong bonds with molecules on other cells and with internal or external fibers.

In this chapter, we provide a framework for understanding the primacy of cells in biological systems and review several fundamental concepts that recur throughout our more detailed discussions in subsequent chapters. We begin with a brief look at the role of evolution and then discuss the general properties of the molecules found in biological systems. Next, we review the main features of cellular architecture, noting the similarities and differences between the three main cell lineages that have emerged over evolutionary time. The remaining topics covered in this chapter focus on the assemblage of cells into organized structures and their dynamic nature, in preparation for later chapters dealing with various processes critical to cellular growth, differentiation, and adaptation to changing circumstances.

EVOLUTION: AT THE CORE OF MOLECULAR CHANGE

The interplay of events played out over billions of years, in the historical process called *evolution,* dictates the form and structure of the living world today. Thus biology, which is the study of the results of these historical events, differs fundamentally from physics and chemistry, which deal with the essential and unchanging properties of matter. The great insight of Charles Darwin was that all organisms are related in a great chain of being extending from the distant past to the present. The Darwinian principle that organisms vary randomly and the fittest are then selected by the forces of their environment guides biological thinking to this day.

We now know that genes, which chemically are composed of deoxyribonucleic acid (DNA), ultimately define biological structure and maintain the integration of cellular function. The genes encode proteins, the primary molecules that make up cell structures and carry out cellular activities. Alterations in the structure and organization of genes thus provide the random variation that nurtures evolutionary change in biological structure and function.

Even scientists brought up in the evolutionary tradition have been surprised to learn in recent years just how closely the genes of different species are related. During evolution, genes have been conserved to such an extent that some human genes will function in a yeast cell and quite a few will function in a fly cell. Clearly, one feature of evolution is the *maintenance* unchanged of many aspects of cellular life even while great changes in external form and capability are occurring. Recent progress in determining the sequences of all the genes in a variety of organisms is revealing the subtle changes that have fueled evolution.

The creative part of the evolutionary process is *adaptation* to rapidly changing environments and the conquest of new environmental niches. During this process, small alterations in cellular structures and functions are selected. Entirely new structures rarely are created; more often, old structures are adapted to new circumstances. More rapid change is possible by rearranging or multiplying previously evolved components rather than by waiting for a wholly new approach to emerge. The cellular organization of organisms plays a fundamental role in this process because it allows change to come about by small alterations in previously evolved cells, giving them new capabilities

THE MOLECULES OF LIFE

Among the many events that occur in the life of a cell are a multitude of specific chemical transformations, which provide the cell with usable energy and the molecules needed to form its structure and coordinate its activities. Here we briefly describe the functions of the main types of chemicals that compose cells. Throughout many later chapters we will focus on the interactions and transformations of these molecules.

Water, inorganic ions, and a large array of relatively small organic molecules (e.g., sugars, vitamins, fatty acids) account for 75 – 80 percent of living matter by weight. Of these small molecules, water is by far the most abundant. The remainder of living matter consists of macromolecules, including proteins, polysaccharides, and DNA. Cells acquire and use these two size classes of molecules in fundamentally different ways. Ions, water, and many small organic molecules are imported into the cell. Cells also make and alter many small organic molecules by a series of different chemical reactions. In contrast, cells can obtain macromolecules only by making them. Their synthesis entails linking together a specific set of small molecules (monomers) to form polymers through repetition of a single type of chemical-linkage reaction.

Some small molecules function as precursors for synthesis of macromolecules, and the cell is careful to provide the appropriate mix of small molecules needed. Small molecules also store and distribute the energy for all cellular processes; they are broken down to extract this chemical energy, as when sugar is degraded to carbon dioxide and water with the release of the energy bound up in the molecule. Other small molecules (e.g., hormones and growth factors) act as signals that direct the activities of cells and nerve cells communicate with one another by releasing and sensing certain small signaling molecules. The powerful effect on our body of a frightening event comes from the instantaneous flooding of the body with a small-molecule hormone that mobilizes the "fight or flight" response.

Macromolecules, though, are the most interesting and characteristic molecules of living systems; in a true sense the evolution of life as we know it is the evolution of macromolecular structures. Proteins, the workhorses of the cell, are the most abundant and functionally versatile of the cellular macromolecules. To appreciate the abundance of protein within a cell, we can estimate the number of protein molecules in a typical eukaryotic cell, such as a hepatocyte in the liver. This cell, roughly a cube 15 μm (0.0015 cm) on a side, has a volume of 3.4×10^{-9} cm^3 (or milliliters). Assuming a cell density of 1.03 g/ml, the cell would weigh 3.5×10^{-9} g. Since protein accounts for approximately 20 percent of a cell's weight, the total weight of cellular protein is 7×10^{-10} g. The average yeast protein has a molecular weight of 52,700 (g/mol). Assuming this value is typical of eukaryotic proteins, we can calculate the total number of protein molecules per liver cell as about 7.9×10^9 from the total protein weight and the number of molecules per mole, which is a constant (Avogadro's number). To carry this calculation one step further, consider that a liver cell contains about 10,000 different proteins; thus, a cell contains close to a million molecules of each protein on average. In actuality, however, the abundance of different proteins varies widely, from the quite rare cell-surface protein that binds the hormone insulin (20,000 molecules) to the abundant structural protein actin (5×10^8 molecules).

Many of the proteins within cells are enzymes, which accelerate (catalyze) reactions involving small molecules. Other proteins allow cells to move and do work, maintain internal cell rigidity, and transport molecules across membranes. Proteins even direct their own synthesis and that of other macromolecules. Reflecting their numerous functions, proteins come in many shapes and sizes. The elucidation of the structure of proteins and the relation of protein structure to function remain active areas of scientific investigation. Proteins are formed from only 20 different monomers, the amino acids. That such a limited set of building blocks can do so much is a continuous marvel, even to researchers who work with proteins every day. They are the true glory of the biological world.

The macromolecule that garners the most public attention is not protein but deoxyribonucleic acid (DNA), whose functional properties make it the cell's master molecule. The three-dimensional structure of DNA, first proposed by James D. Watson and Francis H. C. Crick about 50 years ago, consists of two long helical strands that are coiled around a common axis forming a double helix. The double-helical structure of DNA, one of nature's most magnificent constructions, is critical to the phenomenon of heredity, the transfer of genetically determined characteristics from one generation to the next.

Each strand of DNA is composed of just four different types of monomers called nucleotides. Genes are simply coded representations of the structures of individual proteins, a code written in four chemical "letters" — the nucleotides — and displayed as a continually varying sequence in DNA. Since cells use proteins (enzymes) to make other molecules like sugars or fats, DNA indirectly directs the synthesis of many small molecules as well as proteins. DNA also contains a coded set of instructions about when various proteins are to be made and in what quantities.

In the common view, DNA is the storage form of genetic information, which protein "machines" read out for use by the cell. But a third macromolecule, ribonucleic acid (RNA), is necessary in the process. The *central dogma* of biology states that the coded genetic information hard-wired into DNA is transcribed into individual transportable cassettes, composed of messenger RNA (mRNA); each mRNA cassette contains the programme for synthesis of a particular protein (or small number of proteins). This critical trio of macromolecules — DNA, RNA, and proteins — is present in all cells. The mechanism whereby the information encoded in DNA is deciphered into proteins is now understood quite well and explained. How this process of gene expression is regulated — that is, how cells "know" to make the right proteins at the right time in the right amounts — is a major focus of current research in molecular cell biology and a recurring theme throughout this book.

THE ARCHITECTURE OF CELLS

Although generalizations in biology usually lack the theoretical underpinnings found in physics, there are very clear commonalities among living systems that give biology a unity. One is the style of cellular construction. The biological universe consists of two types of cells — *prokaryotic cells,* which lack a defined nucleus and have a simplified internal organization, and *eukaryotic cells,* which have a more complicated internal structure including a defined, membrane-limited nucleus. Detailed analysis of the DNA from a variety of prokaryotic organisms in recent years has revealed two distinct types: bacteria (often called "true" bacteria or eubacteria) and archaea (also called *archaebacteria* or *archaeans*). As we discuss the archaea are in some respects more similar to eukaryotic organisms than to the true bacteria.

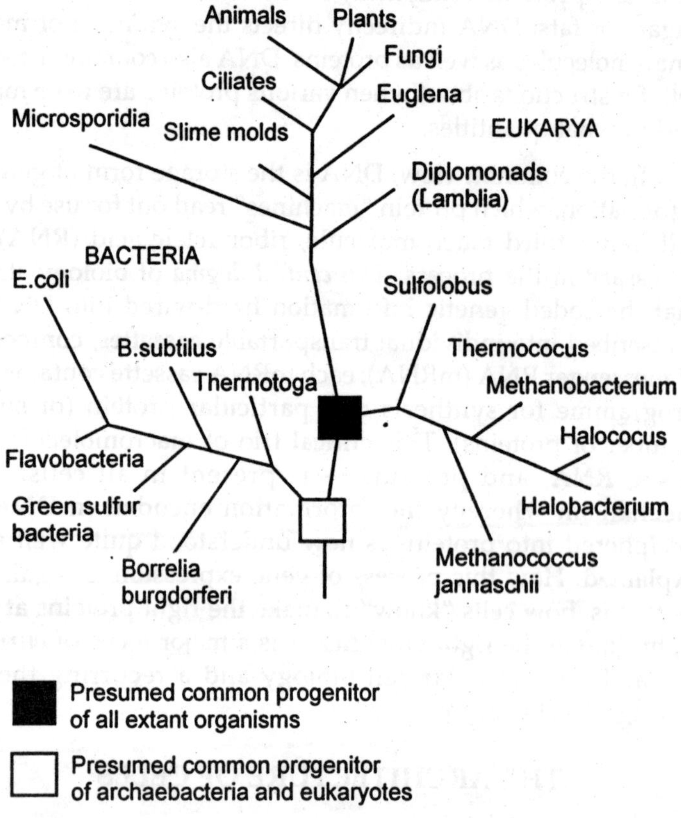

Fig. The three kingdoms of organisms are related through common sequences of their ribosomal RNAs.

Based on the assumption that organisms with more similar genes evolved from a common progenitor more recently than those with more dissimilar genes, researchers have developed the lineage tree. According to this tree, the archaea and eukarya (eukaryotes) are thought to have diverged from the bacteria before they diverged from each other. Despite the differences in the organization of prokaryotic and eukaryotic cells, all cells share certain structural features and carry out many complicated processes in basically the same way.

CELLS ARE SURROUNDED BY WATER-IMPERMEABLE MEMBRANES

A cell, because it is a limited space, must have an outer border. The construction of that border represents one of the most fundamental considerations in biological organization. The outer shell of cells, like any shell, is built to keep the interior contents from leaking out into the surrounding environment. The chemical processes of cellular life generally take place in a watery solution, and the intracellular constituents of cells are largely molecules that are easily dissolved in water. Similarly, the environment around cells is a watery one, the blood and other bodily fluids being solutions in water. Cells then, in order to maintain their integrity, need to be surrounded by an environment through which water cannot flow. A membrane composed of fatty molecules serves this purpose.

We all know from common experience that "oil and water don't mix." That maxim is all one needs to appreciate how a cell is constructed. When oil is poured on water, the oil spreads into a thin film; that film is analogous to the film of fat that surrounds cells, called the plasma membrane. Biological membranes differ from a pure oil film in that the molecules that make the membrane have both oily and watery portions; they have long fatty chains, but they also have a head group that is water-soluble by virtue of being electrically charged. Thus membranes are formed because these bipartite molecules, called phospholipids, spontaneously orient themselves to form a double layer, or bilayer, having a fatty interior with external surfaces bonded to the surrounding water by the charged head groups. The membrane is given rigidity by interspersion of cholesterol, a molecule we have come to hate because of its association with heart disease, but one that is required to build the outer membrane of all our cells. Hence from an understanding of the contrasting properties of watery solutions and oily layers, an understanding of cellular construction emerges.

In spite of the rigidity provided by cholesterol, membranes composed of fat are not very strong, so numerous mechanisms for strengthening the borders of cells have evolved. In plants the plasma membrane is surrounded by a rigid cell wall. Although most animal cells lack a cell wall, proteins attached to their exterior surfaces provide some stability; the linking of cells together through these proteins helps maintain the integrity of tissues. Tissues and organs are often covered by strong networks of proteins and other molecules that strengthen and protect them, and also wall off the various compartments of the body. Single-celled organisms, like bacteria, have special outer coats to protect them.

MEMBRANES SERVE FUNCTIONS OTHER THAN SEGREGATION

Although membranes are valuable as a way to segregate the watery interior of the cell from its environment, or to segregate intracellular events from one another, they have other important functions, including energy storage. Because membranes separate watery compartments from one another, if an ion or a molecule dissolved in water is moved through a membrane into a new cellular compartment, it will not be able to diffuse freely out of the compartment into which it was moved. It takes energy to move the molecule, but once moved, the molecule stores that energy by virtue of its entrapment. Formally, this storage of energy is just like the storage of energy in a battery. Therefore, membranes not only delineate compartments, but also serve as active participants in the cell's dynamism.

The functions of many proteins depend on their mode of association with membranes. For instance, the passage of water-soluble molecules through membranes is carried out by protein transporters that are embedded in the membrane. Also, cells send information to one another by releasing signaling molecules. The outer membranes of cells have proteins, known appropriately as *receptors,* that bind the circulating signaling molecules. These signaling molecules allow the individual activities of the many cells in the body to be coordinated. The receipt of a signaling molecule by a receptor causes the transient organization of particular types of intracellular proteins, called *signal-transduction proteins,* into an activated complex at the interior face of the cell's outer membrane, from which it directs alterations of events in the cell's cytoplasm or nucleus.

PROKARYOTES COMPRISE A SINGLE MEMBRANE LIMITED COMPARTMENT

All prokaryotes are single-celled organisms, or protists. The bacterial lineage includes *Escherichia coli,* found in animal intestines and a favourite experimental organism, and the photosynthetic organisms formerly known as *blue-green algae* but better known today as *cyanobacteria.* (Because most prokaryotes studied in laboratories are bacteria, discussions of prokaryotic structure or metabolism throughout this book refer to these organisms, not archaeans, unless noted otherwise.) Many members of the archaeal lineage grow in unusual, often extreme, environments. For instance, the halophiles require high concentrations of salt to survive, and the thermoacidophiles grow in hot (80°C) sulfur springs, where a pH of less than 2 is common. Other archaeans, called *methanogens,* live in oxygen-free milieus and generate methane (CH_4) by the reduction of carbon dioxide.

Archaeal cells have a similar structure. In general, prokaryotes consist of a single closed compartment containing the cytosol and bounded by the plasma membrane. Although bacterial cells do not have a defined nucleus, the genetic material, DNA, is condensed into the central region of the cell. In addition, most ribosomes — the cell's protein-synthesizing particles — are found in the DNA-free region of the cell. Some bacteria also have an invagination of the cell membrane, called a *mesosome,* which is associated with synthesis of DNA and secretion of proteins. Thus bacterial cells are not completely devoid of internal organization.

Bacterial cells possess a cell wall, which lies adjacent to the external side of the plasma membrane. The cell wall is composed of layers of peptidoglycan, a complex of proteins and oligosaccharides; it helps protect the cell and maintain its shape. Some bacteria (e.g., *E. coli*) have a thin cell wall and an unusual outer membrane separated from the cell wall by the periplasmic space. Such bacteria are not stained by the Gram technique and thus are classified as gram-negative. Other bacteria (e.g., *Bacillus polymyxa*) that have a thicker cell wall and no outer membrane take the Gram stain and thus are classified as gram-positive.

EUKARYOTIC CELLS CONTAIN MANY ORGANELLES AND A COMPLEX CYTOSKELETON

Eukaryotes comprise all members of the plant and animal kingdoms, including the unicellular fungi (e.g., yeasts, mushrooms, molds) and protozoans. Eukaryotic cells, like prokaryotic cells, are surrounded by a plasma membrane. However, unlike prokaryotic cells, most eukaryotic cells also contain extensive internal membranes that enclose specific compartments, the organelles, and separate them from the rest of the cytoplasm, the region of the cell lying outside the nucleus.

Most organelles are surrounded by a single phospholipid membrane, but several, including the nucleus, are enclosed by two membranes. Each type of organelle plays a unique role in the growth and metabolism of the cell, and each contains a collection of specific enzymes that catalyze requisite chemical reactions. The membranes defining these subcellular compartments control their internal ionic composition so that it commonly differs from that of the cytosol (the portion of the cytoplasm outside the organelles) and among the various organelles.

The largest organelle in a eukaryotic cell is generally the nucleus, which houses most of the cellular DNA. In addition to the nucleus, several other organelles are present in nearly all eukaryotic cells: the mitochondria, in which much of the cell's energy metabolism is carried out; the rough and smooth endoplasmic reticula, a network of membranes in which glycoproteins and lipids are synthesized; Golgi vesicles, which direct membrane constituents to appropriate places in the cell; and peroxisomes, in which fatty acids and amino acids are degraded. Animal cells, but not plant cells, contain lysosomes, which degrade worn-out cell constituents and foreign materials taken in by the cell. Chloroplasts, where photosynthesis occurs, are found only in certain leaf cells of plants and some single-celled organisms. Both plant cells and some single-celled eukaryotes contain one or more vacuoles, large, fluid-filled organelles in which nutrients and waste compounds are stored and some degradative reactions occur.

The cytosol of eukaryotic cells contains an array of fibrous proteins collectively called the cytoskeleton. Three classes of fibers compose the cytoskeleton: microtubules (20 nm in diameter), built of polymers of the protein tubulin; microfilaments (7 nm in diameter), built of the protein actin; and intermediate filaments (10 nm in diameter), built of one or more rod-shaped protein subunits. The cytoskeleton gives the cell strength and rigidity, thereby helping to maintain cell shape. Cytoskeletal fibers also control movement of structures within the cell; for example, some cytoskeletal fibers connect to organelles or provide tracks along which organelles move.

The rigid cell wall, composed of cellulose and other polymers, that surrounds plant cells contributes to their strength and rigidity. Fungi are also surrounded by a cell wall, but its composition differs from that of bacterial or plant cell walls.

CELLULAR DNA IS PACKAGED WITHIN CHROMOSOMES

The DNA in the nuclei of eukaryotic cells is distributed among 1 to more than 50 long linear structures called chromosomes. The number and size of the chromosomes are the same in all cells of an organism, but vary among different types of organisms. Each chromosome comprises a single DNA molecule associated with numerous proteins, and the total

DNA in the chromosomes of an organism is referred to as its genome. Chromosomes, which stain intensely with basic dyes, are visible in the light microscope only during cell division when the DNA becomes tightly compacted.

In all prokaryotic cells, most of or all the genetic information resides in a single circular DNA molecule, about a millimeter in length; this molecule lies, folded back on itself many times, in the central region of the cell. Although the large genomic DNA molecule in prokaryotes is associated with proteins and often is referred to as a chromosome, the arrangement of DNA within a bacterial chromosome differs greatly from that within the chromosomes of eukaryotic cells.

The concept that genes are like "beads" strung on a long "string," the chromosome, was proposed early in the 1900s based on genetic work with the fruit fly *Drosophila*. The early *Drosophila* workers could position, or map, the genes responsible for various mutant traits on a chromosome, even though they did not yet know that genes were segments of DNA or that the function of a gene was due to a protein whose sequence was encoded by that gene!

THE LIFE CYCLE OF CELLS

A cell in an adult organism can be viewed as a steady-state system. The DNA is constantly read out into a particular set of mRNAs, which specify a particular set of proteins. As these proteins function, they are also being degraded and replaced by new ones, and the system is so balanced that the cell neither grows, shrinks, nor changes its function. This static view of the cell, however, misses the all-important dynamic aspects of cellular life.

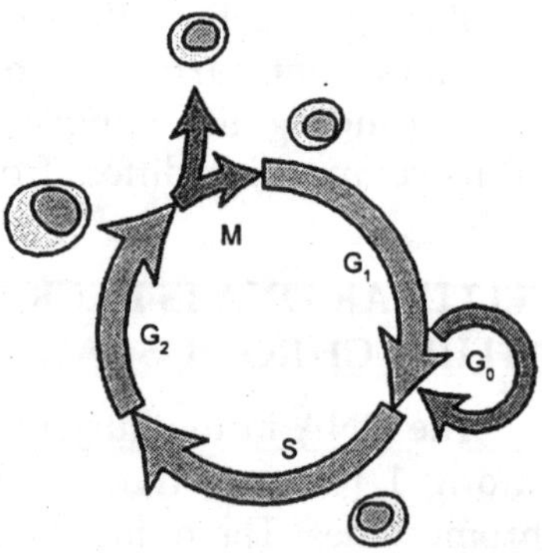

Fig. The eukaryotic cell cycle

The dynamics of a cell can best be understood by examining the course of its life. A new cell arises when one cell divides or when two cells, like a sperm and an egg cell, fuse. Either event sets off a cell-replication programme that is encoded in the DNA and executed by proteins. This programme usually involves a period of cell growth, during which proteins are made and DNA is replicated, followed by cell division, when a cell divides into two daughter cells. Whether a given cell will grow and divide is a highly regulated decision of the body, assuring that an adult organism replaces worn out cells or makes more cells in response to a new need. Examples of the latter are the growth of muscle in response to exercise or damage, and the proliferation of red blood cells when a person ascends to a higher altitude and needs more capacity to capture oxygen. However, in one major and devastating disease — cancer — cells multiply even though they are not needed by the body. To understand how cells become cancerous, biologists have intensely studied the mechanisms that control the growth and division of cells.

THE CELL CYCLE FOLLOWS A REGULAR TIMING MECHANISM

Most eukaryotic cells live according to an internal clock; that is, they proceed through a sequence of phases, called the cell cycle, during which DNA is duplicated during the synthesis (S) phase and the copies are distributed to opposite ends of the cell during mitotic (M) phase. Progress along the cycle is controlled at key checkpoints, which monitor the status of a cell, for instance, the internal amount of DNA or the presence of extracellular nutrients. When certain conditions are met, the cell proceeds to the next checkpoint. The cycle begins after the cell divides into two daughter cells, each containing an identical copy of the parental cell's genetic material.

The cell cycle of prokaryotes is simple and fast. Replication of the single chromosome begins at a particular DNA sequence, the replication origin, which is anchored to the cell membrane. Once DNA replication is complete, assembly of new membrane and cell wall forms a septum, which eventually divides the cell in two. Because the origins of the two newly formed chromosomes are anchored to different membrane sites, each daughter cell receives one chromosome. In ideal growth conditions, the bacterial cell cycle is repeated every 30 minutes.

Only a few types of eukaryotic cells can grow and divide as quickly as bacteria. Most growing plant and animal cells take 10 – 20 hours to double in number, and some duplicate at a much slower rate. Many cells in adult animals, such as nerve cells and striated muscle cells, do not divide at all. They have temporarily exited from the cell cycle after mitosis and entered a "paused or quiescent" state called G_0. Because eukaryotic cells are larger and more complex than prokaryotic cells, a specialized mechanism coordinates their replication of genomic DNA, distribution of chromosomes, and cell division.

MITOSIS APPORTIONS THE DUPLICATED CHROMOSOMES EQUALLY TO DAUGHTER CELLS

Mitosis is the mechanism in eukaryotes for partitioning the genome equally at cell division. To accomplish this complex task, plant and animal cells build a specialized machine, called the mitotic apparatus, which captures the chromosomes and then pushes and pulls them to opposite sides of the dividing cell. Remarkably, the mitotic apparatus is a temporary structure that exists only during mitosis to distribute the genetic material. Although the events of mitosis unfold continuously, they are conventionally divided into four substages representing phases of chromosome movement. During the first substage, prophase, the replicated chromosomes, each comprising two identical chromatids, are condensed into compact packets and then released to the cytoplasm when the nuclear membrane breaks down. During metaphase and anaphase, the chromosomes are sorted, and each chromatid of a pair moves to opposite sides of the cell. The end of mitosis is marked by re-formation of a membrane around each set of chromosomes (telophase). Division of the cytoplasm, called cytokinesis, then yields two daughter cells, each with a $2n$ complement of genetic material.

Cell division in plant and animal cells differs mainly at cytokinesis. Animal cells divide in two by pinching of the cytoplasm. However, because a plant cell is surrounded by a rigid cell wall, daughter cells are formed by building a new cell membrane and cell wall between the two daughter nuclei, thereby cutting the cytoplasm into two portions.

CELL DIFFERENTIATION CREATES NEW TYPES OF CELLS

The most complicated example of cellular dynamics occurs when a cell changes, or *differentiates,* to carry out a specialized function. This process often is marked by a change in the microscopic appearance, or *morphology,* of the cell. For example, the different structures of a nerve cell and a muscle cell reflect their respective functions in long-distance communication and contraction, highlighting the biological principle that "form follows function."

Cell differentiation creates the diversity of cell types that arise during the development of an organism from a fertilized egg. This is a process of extensive cell multiplication and differentiation. A mammal that starts as one cell becomes an organism with hundreds of diverse cell types such as muscle, nerve, and skin. Here we see at its most dramatic the power of DNA to control cellular behavior: development is a DNA-orchestrated set of cellular changes (easily tens of thousands of them) that occur virtually without fail. The almost perfect resemblance of "identical" twins is a testament to the programme encoded by DNA to reproducibly direct the development of a human being.

Nowhere is the variety of cellular activities and responses better illustrated than in the body's immune system. It is there that many cell types come together in organized tissues specifically designed to allow the body to distinguish its own cells from those of foreign invaders. Within the immune system, we see both development of specialized cells that can recognize invading cells and formation of tissues from cells that originate in various parts of the body. The immune-system cells not only actively survey their environment with surface receptor proteins like antibodies, but also change their properties when they encounter a foreign substance, allowing the body to rid itself of invaders.

CELLS DIE BY SUICIDE

Unchecked cell growth and multiplication produce a mass of cells, a tumor. *Programmemed cell death* plays the very important role of population control by balancing cell growth and multiplication. In addition, cell death also eliminates unnecessary cells. For example, during embryogenesis, the digits of our fingers and toes are sculpted by the death of cells in the intervening spaces. If these cells remained alive, our hands and feet would become webbed. Thus the timing and location of cell death, as well as cell growth and division, must be precisely controlled.

Cell death follows an internal programme of events called apoptosis, in which all traces of a cell vanish. The first visible sign of apoptosis is condensation of the nucleus and fragmentation of the DNA. The cell soon shrivels and is consumed by macrophages. A cell is directed to commit suicide when an essential factor is removed from the extracellular environment or when an internal signal is activated. Thus, the default state of the cell is to remain alive. The discovery of genes that suppress the growth of tumors by activating cell death stimulated an exciting new line of cancer research that may lead to more effective treatment strategies

INTEGRATING CELLS INTO TISSUES

The evolution of multicellular organisms permitted specialized cells and tissues to form; a flowering plant has at least 15 cell types, and a vertebrate hundreds. In both plants and animals, cells that are specialized to carry out a particular task are found together in the tissues in which the task is performed: a xylem or meristem; a liver, a muscle, or a nerve ganglion. Different types of cells in a tissue are often arranged in precise patterns of staggering complexity. For instance, the hundreds of different types of neurons in the human brain are interconnected to one another through a network of some 10^{15} synaptic connections! The coordinated functioning of many types of cells within tissues, and of multiple specialized tissues, permits the organism as a whole to move, metabolize, reproduce, and carry out other essential activities.

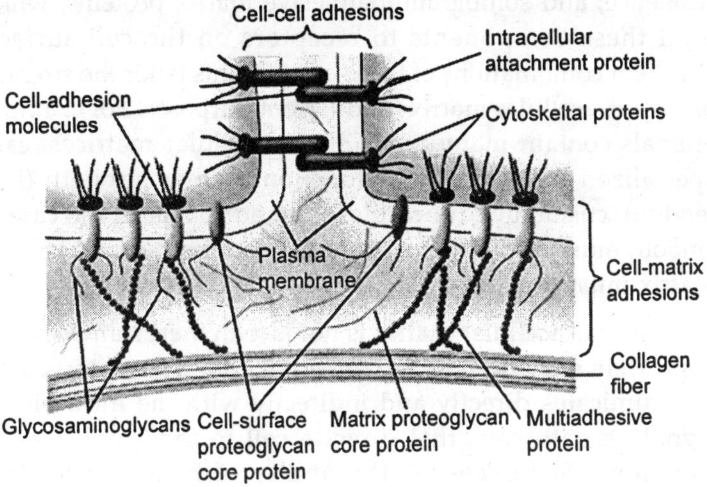

Fig. Integrating Cells into Tissues.

A key step in the evolution of multicellularity must have been the ability of cells to contact tightly and interact specifically with other cells. Various integral membrane proteins, collectively termed cell-adhesion molecules (CAMs), enable many animal cells to adhere tightly and specifically with cells of the same, or similar, type; these interactions allow populations of cells to segregate into distinct tissues. Following aggregation, cells elaborate specialized cell junctions that stabilize these interactions and promote local communication between adjacent cells. Animal cells also secrete a complex network of proteins and carbohydrates, the extracellular matrix (ECM), that creates a special environment in the spaces between cells. The matrix helps bind the cells in tissues together and is a reservoir for many hormones controlling cell growth and differentiation. The matrix also provides a lattice through which cells can move, particularly during the early stages of differentiation. Defects in these connections lead to cancer and developmental malformations.

The extracellular matrix has three major protein components: highly viscous proteoglycans, which cushion cells; insoluble collagen fibers, which provide strength and resilience; and soluble multiadhesive matrix proteins, which bind these components to receptors on the cell surface. Different combinations of these components tailor the strength of the extracellular matrix for different purposes. For example, animals contain many types of extracellular matrices, each specialized for a particular function such as strength (in a tendon), cushioning (in cartilage), or adhesion. In the case of smooth muscle cells that surround an artery, the extracellular matrix must provide strong but flexible connections.

The extracellular matrix is not just an inert framework or cage that supports or surrounds cells. The matrix also communicates directly and indirectly with the intracellular signaling pathways that direct a cell to carry out specific functions. For example, the ability of hepatocytes—the principal cells in the liver—to express liver-specific proteins depends on their association with a matrix of appropriate composition. Specific ECM components can directly activate cytosolic signal-transduction pathways by binding to cell-adhesion protein receptors in the plasma membrane. Alternatively, by binding growth factors and other hormones, the matrix can either sequester these signals from cells or, conversely, present them to cells, thereby indirectly inducing or inhibiting signaling pathways. Morphogenesis—the later stage of embryonic development during which form is achieved by cell movements and rearrangements—also is critically dependent on ECM components, which are constantly being remodeled, degraded, and resynthesized locally. Even in adults—in areas of wounding, for example — degradation and resynthesis of ECM components occurs.

In this chapter, we focus on the structure of and interactions between ECM components, cell-adhesion molecules, and cell-adhesion junctions—the main structures that permit animal cells to form organized tissues. Plant cells are surrounded by a cell wall that is thicker and more rigid than the extracellular matrix. Although the plant cell wall and the extracellular matrix serve many of the same functions, they are structurally very different. Because of these differences, we discuss the plant cell wall and its interactions with plant cells in a separate section at the end of the chapter.

Chapter 2

The Chemical Foundations

In this chapter we review many important chemical concepts required to comprehend cellular processes, all of which follow the rules of chemistry. Constituting 70 – 80 percent by weight of most cells, water is the most abundant of the "chemicals of life." About 7 percent of the weight of living matter is composed of inorganic ions and small molecules such as nucleotides (the building blocks of DNA and RNA), amino acids (the building blocks of proteins), and sugars. All these small molecules can be chemically synthesized in the laboratory.

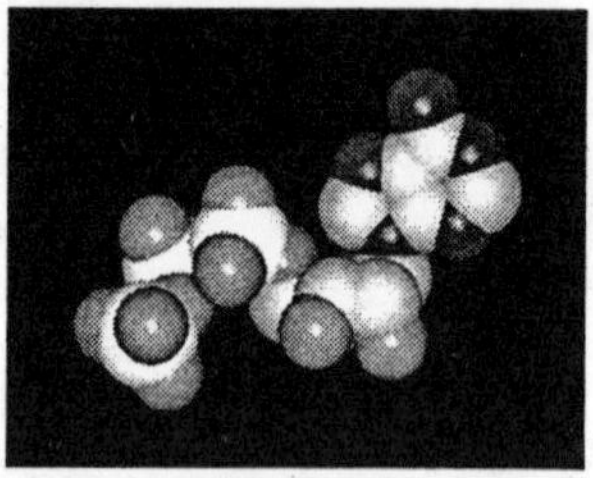

Figure. Three-dimensional structure of an ATP molecule.

The principal cellular macromolecules — DNA, RNA, and protein — compose the remainder of living matter. Like the small molecules found in cells, these very large molecules follow the general rules of chemistry and can be chemically synthesized. Molecular cell biology, however, aims at understanding higher levels of biological organization, that is, explaining the structure and function of organisms and cells in terms of the properties of individual molecules, such as proteins and nucleic acids. In this book, we examine many examples of complex, higher-order processes that are understood, at least partially, at the molecular level. The realization that complex processes such as evolution, development of an organism from a fertilized egg, motion, perception, and thought follow the rules of chemistry and physics — and that no vital or supernatural force is involved — has profound philosophical and even political implications.

All the topics discussed in this chapter relate to important concepts. We begin with a discussion of covalent bonds, which connect individual atoms in a molecule, and the structures of carbohydrates, which illustrate the importance of relatively small differences in the arrangements of covalent bonds. Next we consider noncovalent bonds, important stabilizing forces between groups of atoms within larger molecules and between different molecules. The four most important ones are hydrogen bonds, ionic interactions, van der Waals interactions, and hydrophobic bonds. We discuss the role of hydrogen bonds in determining the properties of water, the major constituent of cells and of the spaces between cells. Phospholipids and their association into a bilayer structure, stabilized by multiple noncovalent bonds, are also described. The phospholipid bilayer is the basis of all biological membranes (biomembranes), which have numerous functions discussed in later chapters. The next section reviews the concept of chemical equilibrium, focusing on the properties of chemicals in aqueous solution; in particular, the pH of a solution, the reactions of acids and bases, and the buffering capacity of solutions. We then consider various aspects of biochemical energetics, including the central concept of free energy, its relationship to the direction of chemical reactions, and the central role of ATP (adenosine triphosphate) in capturing and transferring energy in cellular metabolism. In the final section, we describe the factors that control the rates of chemical reactions and how enzymes accelerate reaction rates.

COVALENT BONDS

Covalent bonds, which hold the atoms within an individual molecule together, are formed by the sharing of electrons in the outer atomic orbitals. The distribution of shared as well as unshared electrons in outer orbitals is a major determinant of the three-dimensional shape and chemical reactivity of molecules. For instance, as we learn the shape of proteins is crucial to their function and their interactions with small molecules. In this section, we discuss important properties of covalent bonds and describe the structure of carbohydrates to illustrate how the geometry of bonds determines the shape of small biological molecules.

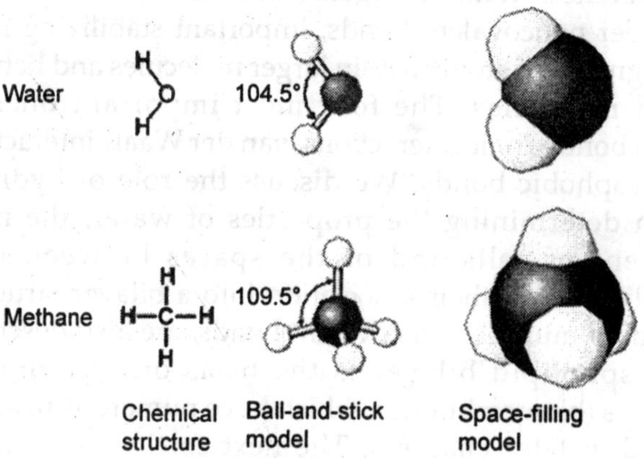

Fig.Covalent bonds

EACH ATOM CAN MAKE A DEFINED NUMBER OF COVALENT BONDS

Electrons move around the nucleus of an atom in clouds called *orbitals,* which lie in a series of concentric *shells,* or energy levels; electrons in outer shells have more energy than those in inner shells. Each shell has a maximum number of electrons that it can hold. Electrons fill the innermost shells of an atom first; then the outer shells. The energy level of an atom is lowest when all of its orbitals are filled, and an atom's reactivity depends on how many electrons it needs to complete its outermost orbital. In most cases, in order to fill the outermost orbital, the electrons within it form covalent bonds with other atoms. A covalent bond thus holds two atoms close together because electrons in their outermost orbitals are shared by both atoms.

Most of the molecules in living systems contain only six different atoms: hydrogen, carbon, nitrogen, phosphorus, oxygen, and sulfur. The outermost orbital of each atom has a characteristic number of electrons:

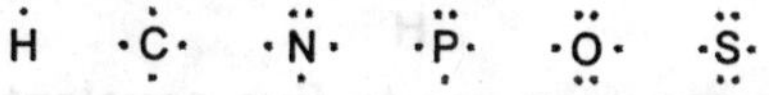

These atoms readily form covalent bonds with other atoms and rarely exist as isolated entities. As a rule, each type of atom forms a characteristic number of covalent bonds with other atoms.

For example, a hydrogen atom, with one electron in its outer shell, forms only one bond, such that its outermost orbital becomes filled with two electrons. A carbon atom has four electrons in its outermost orbitals; it usually forms four bonds, as in methane (CH_4), in order to fill its outermost orbital with eight electrons. The single bonds in methane that connect the carbon atom with each hydrogen atom contain two shared electrons, one donated from the C and the other from the H, and the outer (*s*) orbital of each H atom is filled by the two shared electrons:

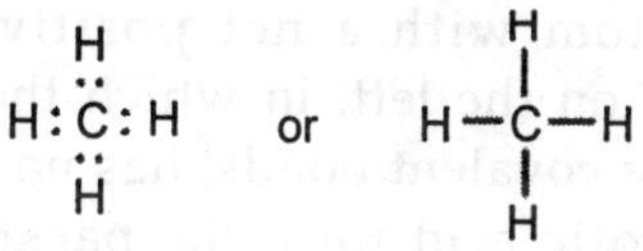

Nitrogen and phosphorus each have five electrons in their outer shells, which can hold up to eight electrons. Nitrogen atoms can form up to four covalent bonds. In ammonia (NH_3), the nitrogen atom forms three covalent bonds; one pair of electrons around the atom (the two dots on the right) are in an orbital not involved in a covalent bond:

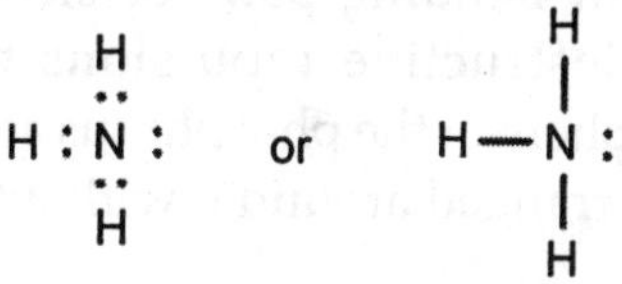

In the ammonium ion (NH_4^+), the nitrogen atom forms four covalent bonds, again filling the outermost orbital with eight electrons:

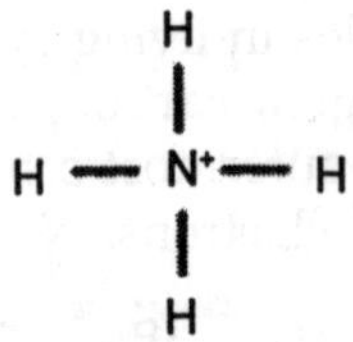

Phosphorus can form up to five covalent bonds, as in phosphoric acid (H_3PO_4). The H_3PO_4 molecule is actually a "resonance hybrid," a structure between the two forms shown below in which nonbonding electrons are shown as pairs of dots:

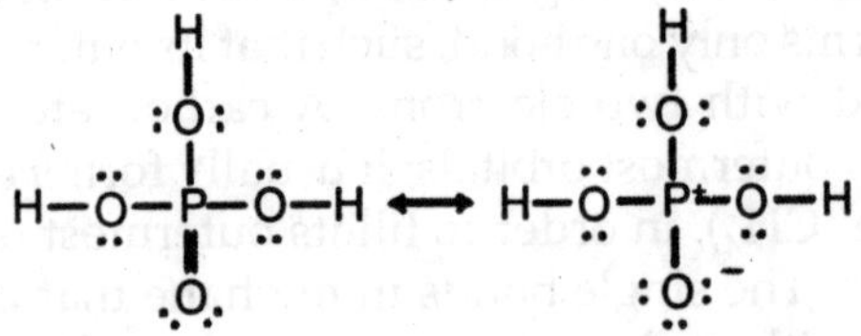

In the resonance hybrid on the right, one of the electrons from the P=O double bond has accumulated around the O atom, giving it a net negative charge and leaving the P atom with a net positive charge. The resonance hybrid on the left, in which the P atom forms the maximum five covalent bonds, has no charged atoms. Esters of phosphoric acid form the backbone of nucleic acids phosphates also play key roles in cellular energetics and in the regulation of cell function.

The difference between the bonding patterns of nitrogen and phosphorus is primarily due to the relative sizes of the two atoms: the smaller nitrogen atom has only enough space to accommodate four bonding pairs of electrons around it without creating destructive repulsions between them, whereas the larger sphere of the phosphorus atom allows more electron pairs to be arranged around it without the pairs being too close together.

Both oxygen and sulfur contain six electrons in their outermost orbitals. However, an atom of oxygen usually forms only two covalent bonds, as in molecular oxygen, O_2:

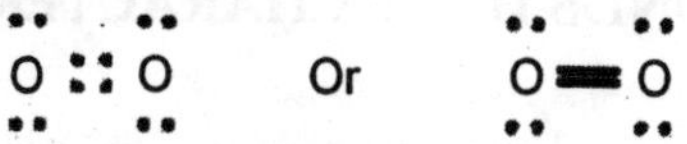

Primarily because its outermost orbital is larger than that of oxygen, sulfur can form as few as two covalent bonds, as in hydrogen sulfide (H_2S), or as many as six, as in sulfur trioxide (SO_3) or sulfuric acid (H_2SO_4):

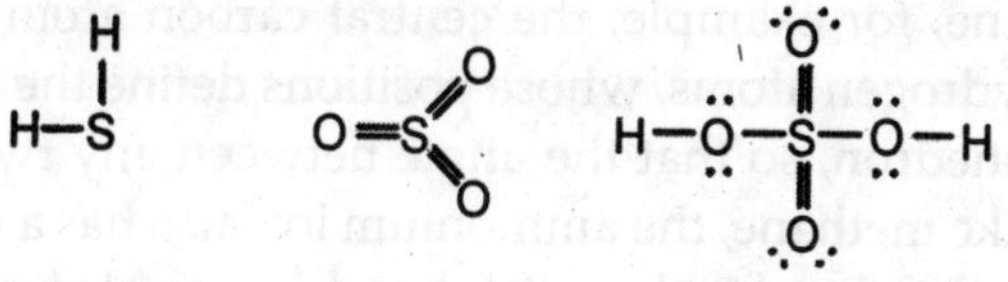

Hydrogen sulfide Sulfur trioxide Sulfuric acid

Esters of sulfuric acid are important constituents of the proteoglycans that compose part of the extracellular matrix surrounding most animal cells.

THE MAKING OR BREAKING OF COVALENT BONDS INVOLVES LARGE ENERGY CHANGES

Covalent bonds tend to be very stable because the energies required to break or rearrange them are much greater than the thermal energy available at room temperature (25 °C) or body temperature (37 °C). For example, the thermal energy at 25 °C is less than 1 kilocalorie per mole (kcal/mol), whereas the energy required to break a C—C bond in ethane is about 83 kcal/mol:

$$H_3C : CH_3 \longrightarrow H_3C\cdot + \cdot CH_3 \quad \Delta H = +83 \text{kcl/mol}$$

where ΔH represents the difference in the total energy of all of the bonds (the enthalpy) in the reactants and in the products.*The positive value indicates that an input of energy is needed to cause the reaction, and that the products contain more energy than the reactants. The high energy needed for breakage of the ethane bond means that at room temperature (25 °C) well under 1 in 10^{12} ethane molecules exists as a pair of $\cdot CH_3$ radicals. The covalent bonds in biological molecules have ΔH values similar to that of the C—C bond in ethane.

COVALENT BONDS HAVE CHARACTERISTIC GEOMETRIES

When two or more atoms form covalent bonds with another central atom, these bonds are oriented at precise angles to one another. The angles are determined by the mutual repulsion of the outer electron orbitals of the central atom. These bond angles give each molecule its characteristic shape. In methane, for example, the central carbon atom is bonded to four hydrogen atoms, whose positions define the four points of a tetrahedron, so that the angle between any two bonds is 109.5°. Like methane, the ammonium ion also has a tetrahedral shape. In these molecules, each bond is a *single bond,* a single pair of electrons shared between two atoms. When two atoms share two pairs of electrons — for example, when a carbon atom is linked to only three other atoms — the bond is a *double bond*:

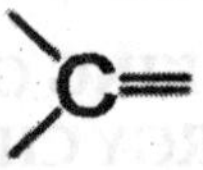

In this case, the carbon atom and all three atoms linked to it lie in the same plane. Atoms connected by a double bond cannot rotate freely about the bond axis, while those in a single bond generally can. The rigid planarity imposed by double bonds has enormous significance for the shape of large biological molecules such as proteins and nucleic acids. (In *triple bonds,* two atoms share six electrons. These are rare in biological molecules.)

All outer electron orbitals, whether or not they are involved in covalent bond formation, contribute to the properties of a molecule, in particular to its shape. For example, the outer shell of the oxygen atom in a water molecule has two pairs of nonbonding electrons; the two pairs of electrons in the H—O bonds and the two pairs of nonbonding electrons form an almost perfect tetrahedron. However, the orbitals of the nonbonding electrons have a high electron density and thus tend to repel each other, compressing the angle between the covalent H—O—H bonds to 104.5° rather than the 109.5° in a tetrahedron.

ELECTRONS ARE SHARED UNEQUALLY IN POLAR COVALENT BONDS

In a covalent bond, one or more pairs of electrons are shared between two atoms. In certain cases, the bonded atoms exert different attractions for the electrons of the bond, resulting in unequal sharing of the electrons. The power of an atom in a molecule to attract electrons to itself, called *electronegativity,* is measured on a scale from 4.0 (for fluorine, the most electronegative atom) to a hypothetical zero. Knowing the electronegativity of two atoms allows us to predict whether a covalent bond can form between them; if the differences in electronegativity are considerable — as in sodium and chloride — an ionic bond, rather than a covalent bond, will form. This type of interaction is discussed in a later section.

In a covalent bond in which the atoms either are identical or have the same electronegativity, the bonding electrons are shared equally. Such a bond is said to be nonpolar. This is the case for C—C and C—H bonds. However, if two atoms differ in electronegativity, the bond is said to be polar. One end of a polar bond has a partial negative charge (δ^-), and the other end has a partial positive charge (δ^+). In an O—H bond, for example, the oxygen atom, with an electronegativity of 3.4, attracts the bondd electrons more than does the hydrogen atom, which has an electronegativity of 2.2. As a result, the bonding electrons spend more time around the oxygen atom than around the hydrogen. Thus the O—H bond possesses an *electric dipole,* a positive charge separated from an equal but opposite negative charge. We can think of the oxygen atom of the O—H bond as having, on average, a charge of 25 percent of an electron, with the H atom having an equivalent positive charge. The *dipole moment* of the O—H bond is a function of the size of the positive or negative charge and the distance separating the charges.

In a water molecule both hydrogen atoms are on the same side of the oxygen atom. As a result, the side of the molecule with the two H atoms has a slight net positive charge, whereas the other side has a slight net negative charge. Because of this separation of positive and negative charges, the entire molecule has a net dipole moment. Some molecules, such as the linear molecule CO_2, have two polar bonds:

$$O^{\delta-} = O^{\delta+} = O^{\delta-}$$

Because the dipole moments of the two C=O bonds point in opposite directions, they cancel each other out, resulting in a molecule without a net dipole moment.

ASYMMETRIC CARBON ATOMS ARE PRESENT IN MOST BIOLOGICAL MOLECULES

A carbon (or any other) atom bonded to four dissimilar atoms or groups is said to be asymmetric. The bonds formed by an asymmetric carbon atom can be arranged in threedimensional space in two different ways, producing molecules that are mirror images of each other. Such molecules are called *optical isomers,* or stereoisomers. One isomer is said to be right-handed and the other left-handed, a property called *chirality*. Most molecules in cells contain at least one asymmetric carbon atom, often called a *chiral carbon* atom. The different stereoisomers of a molecule usually have completely different biological activities.

AMINO ACIDS

Except for glycine, all amino acids, the building blocks of the proteins, have one chiral carbon atom, called the α *carbon,* or C_α, which is bonded to four different atoms or groups of atoms. In the amino acid alanine, for instance, this carbon atom is bonded to $-NH_2$, $-COOH$, $-H$, and $-CH_3$. By convention, the two mirror-image structures are called the D (*dextro*) and the L (*levo*) isomers of the amino acid. The two isomers cannot be interconverted without breaking a chemical bond. With rare exceptions, only the L forms of amino acids are found in proteins.

CARBOHYDRATES

The three-dimensional structures of carbohydrates provide another excellent example of the structural and biological importance of chiral carbon atoms, even in simple molecules. A carbohydrate is constructed of carbon (*carbo-*) plus hydrogen and oxygen (*-hydrate,* or water). The formula for the simplest carbohydrates — the monosaccharides, or simple sugars — is $(CH_2O)_n$, where *n* equals 3, 4, 5, 6, or 7. All monosaccharides contain hydroxyl (—OH) groups and either an aldehyde or a keto group:

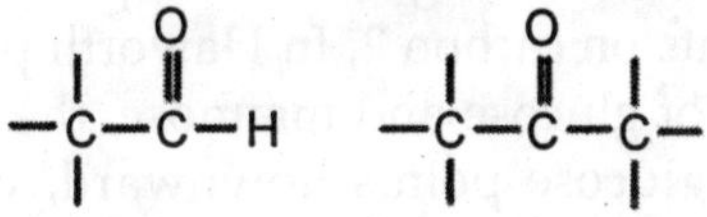

In the linear form of D-*glucose* ($C_6H_{12}O_6$), the principal source of energy for most cells in higher organisms, carbon atoms 2, 3, 4, and 5 are asymmetric. If the hydrogen atom and the hydroxyl group attached to carbon atom 2 (C_2) were interchanged, the resulting molecule would be a different sugar, D-mannose, and could not be converted to glucose without breaking and making covalent bonds. Enzymes can distinguish between this single point of difference.

D-Glucose can exist in three different forms: a linear structure and two different hemiacetal ring structures. If the aldehyde group on carbon 1 reacts with the hydroxyl group on carbon 5, the resulting hemiacetal, D-glucopyranose, contains a six-member ring. Similarly, condensation of the hydroxyl group on carbon 4 with the aldehyde group results in the formation of D-glucofuranose, a hemiacetal containing a five-member ring. Although all three forms of D-glucose exist in biological systems, the pyranose form is by far the most abundant.

When a linear molecule of D-glucose forms a pyranose ring, carbon 1 becomes asymmetric, so two stereoisomers (called *anomers*) of D-glucopyranose are possible. The hydroxyl group attached to carbon 1 "points" down (below the plane of projection) in α-D-glucopyranose and points up (above the plane of projection) in the β anomer. In aqueous solution the α and β anomers readily interconvert spontaneously; at equilibrium there is about one-third α anomer and two-thirds β, with very little of the open-chain form. Because enzymes can distinguish between the α and β anomers of D-glucose, these forms have specific biological roles.

Most biologically important sugars are six-carbon sugars, or hexoses, that are structurally related to D-glucose. Mannose, as noted, is identical with glucose except for the orientation of the substituents on carbon 2. In Haworth projections of the pyranose forms of glucose and mannose, the hydroxyl group on carbon 2 of glucose points downward, whereas that on mannose points upward. Similarly, galactose, another hexose, differs from glucose only in the orientation of the hydroxyl group on carbon 4.

The Haworth projection is an oversimplification be-cause the actual pyranose ring is not planar. Rather, sugar molecules adopt a conformation in which each of the ring carbons is at the centre of a tetrahedron, just like the carbon in methane. The preferred conformation of pyranose structures is the chair. In this conformation, the bonds going from a ring carbon to nonring atoms may take two directions: axial (perpendicular to the ring) and equatorial (in the plane of the ring).

The L isomers of sugars are virtually unknown in biological systems except for L-fucose. One of the unsolved mysteries of molecular evolution is why only D isomers of sugars and L isomers of amino acids were utilized, and not the chemically equivalent L sugars and D amino acids

α AND *β* GLYCOSIDIC BONDS LINK MONOSACCHARIDES

In addition to the monosaccharides discussed above, two common disaccharides, lactose and sucrose, occur naturally. A disaccharide consists of two monosaccharides linked together by a C—O—C bridge called a glycosidic bond. The disaccharide lactose is the major sugar in milk; sucrose is a principal product of plant photosynthesis and is refined into common table sugar.

In the formation of any glycosidic bond, the carbon 1 atom of one sugar molecule reacts with a hydroxyl group of another. As in the formation of most biopolymers, the linkage is accompanied by the loss of water. In principle, a large number of different glycosidic bonds can be formed between two sugar residues. Glucose could be bonded to fructose, for example, by any of the following linkages: $\alpha(1 \rightarrow 1)$, $\alpha(1 \rightarrow 2)$, $\alpha(1 \rightarrow 3)$, $\alpha(1 \rightarrow 4)$, $\alpha(1 \rightarrow 6)$, $\beta(1 \rightarrow 1)$, $\beta(1 \rightarrow 2)$, $\beta(1 \rightarrow 3)$, $\beta(1 \rightarrow 4)$, or $\beta(1 \rightarrow 6)$, where α or β specifies the conformation at carbon 1 in glucose and the number following the arrow indicates the fructose carbon to which the glucose is bound. Only the $\alpha(1 \rightarrow 2)$ linkage occurs in sucrose because of the specificity of the enzyme (the biological catalyst) for the linking reaction.

Glycosidic linkages also join chains of monosaccharides into longer polymers, called polysaccharides, some of which function as reservoirs for glucose. The most common storage carbohydrate in animal cells is glycogen, a very long, highly branched polymer of glucose units linked together mainly by $\alpha(1 \rightarrow 4)$ glycosidic bonds. As much as 10 percent by weight of the liver can be glycogen. The primary storage carbohydrate in plant cells, starch, also is a glucose polymer with $\alpha(1 \rightarrow 4)$ linkages. It occurs in two forms, amylose, which is unbranched, and amylopectin, which has some branches. In contrast to glycogen and starch, some polysaccharides, such as cellulose, have structural and other nonstorage functions. An unbranched polymer of glucose linked together by $\beta(1 \rightarrow 4)$ glycosidic bonds, cellulose is the major constituent of plant cell walls and is the most abundant organic chemical on earth. Because of the different linkages between the glucose units, cellulose forms long rods, whereas glycogen and starch form coiled helices. Human digestive enzymes can hydrolyze $\alpha(1 \rightarrow 4)$ glycosidic bonds, but not $\beta(1 \rightarrow 4)$ bonds, between glucose units; for this reason humans can digest starch but not cellulose. The synthesis and utilization of these polysaccharides are described in later chapters.

NONCOVALENT BONDS

Carbohydrates illustrate the importance of subtle differences in covalent bonds in generating molecules with different biological activities. However, several types of noncovalent bonds are critical in maintaining the three-dimensional structures of large molecules such as proteins and nucleic acids. Noncovalent bonds also enable one large molecule to bind specifically but transiently to another, making them the basis of many dynamic biological processes.

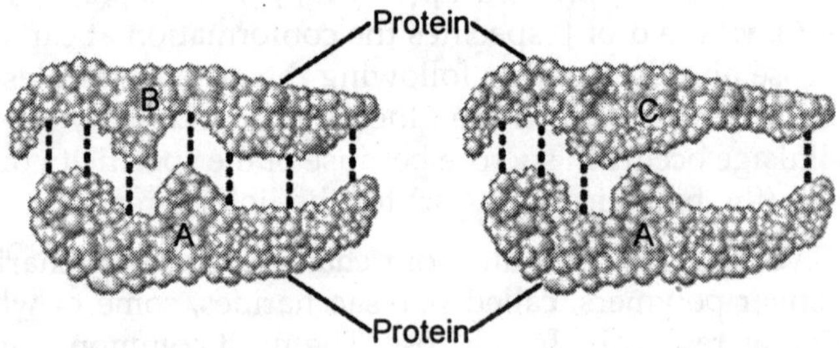

Fig. Multiple weak bonds stabilize specific associations between large molecules.

The energy released in the formation of noncovalent bonds is only 1 – 5 kcal/mol, much less than the bond energies of single covalent bonds. Because the average kinetic energy of molecules at room temperature (25 °C) is about 0.6 kcal/mol, many molecules will have enough energy to break noncovalent bonds. Indeed, these weak bonds sometimes are referred to as *interactions* rather than bonds. Although noncovalent bonds are weak and have a transient existence at physiological temperatures (25 –37 °C), multiple noncovalent bonds often act together to produce highly stable and specific associations between different parts of a large molecule or between different macromolecules. In this section we consider the four main types of noncovalent bonds and discuss their role in stabilizing the structure of biomembranes.

THE HYDROGEN BOND UNDERLIES WATER'S CHEMICAL AND BIOLOGICAL PROPERTIES

Hydrogen bonding between water molecules is of crucial importance because all life requires an aqueous environment and water constitutes about 70–80 percent of the weight of most cells. The mutual attraction of its molecules causes water to have melting and boiling points at least 100 °C higher than they would be if water were nonpolar; in the absence of these intermolecular attractions, water on earth would exist primarily as a gas. The exact structure of liquid water is still unknown. It is believed to contain many transient, maximally hydrogen-bonded networks. Most likely, water molecules are in rapid motion, constantly making and breaking hydrogen bonds with adjacent molecules. As the temperature of water increases toward 100 °C, the kinetic energy of its molecules becomes greater than the energy of the hydrogen bonds connecting them, and the gaseous form of water appears.

PROPERTIES OF HYDROGEN BONDS

Normally, a hydrogen atom forms a covalent bond with only one other atom. However, a hydrogen atom covalently bonded to a donor atom, D, may form an additional weak association, the hydrogen bond, with an acceptor atom, A:

$$D^{\delta-}-H^{\delta+} + :A^{\delta-} \rightleftharpoons D^{\delta-}-\underbrace{H^{\delta+}\cdots\cdots:A^{\delta-}}_{\text{Hydrogen bond}}$$

In order for a hydrogen bond to form, the donor atom must be electronegative, so that the covalent D—H bond is polar. The acceptor atom also must be electronegative, and its outer shell must have at least one nonbonding pair of electrons that attracts the ä$^{+}$ charge of the hydrogen atom. In biological systems, both donors and acceptors are usually nitrogen or oxygen atoms, especially those atoms in amino (—NH_2) and hydroxyl (—OH) groups. Because all covalent N—H and O—H bonds are polar, their H atoms can participate in hydrogen bonds. By contrast, C—H bonds are nonpolar, so these H atoms are almost never involved in a hydrogen bond.

Water molecules provide a classic example of hydrogen bonding. The hydrogen atom in one water molecule is attracted to a pair of electrons in the outer shell of an oxygen atom in an adjacent molecule. Not only do water molecules hydrogen-bond with one another, they also form hydrogen bonds with other kinds of molecules. The presence of hydroxyl (—OH) or amino (—NH_2) groups makes many molecules soluble in water. For instance, the hydroxyl group in methanol (CH_3OH) and the amino group in methylamine (CH_3NH_2) can form several hydrogen bonds with water, enabling the molecules to dissolve in water to high concentrations. In general, molecules with polar bonds that easily form hydrogen bonds with water can dissolve in water and are said to be hydrophilic (Greek, "water-loving"). Besides the hydroxyl and amino groups, peptide and ester bonds are important chemical groups that interact well with water:

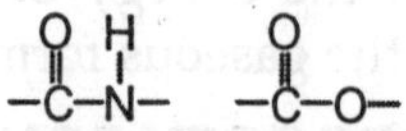

Most hydrogen bonds are 0.26 – 0.31 nm long, about twice the length of covalent bonds between the same atoms. In particular, the distance between the nuclei of the hydrogen and oxygen atoms of adjacent hydrogen-bonded molecules in water is approximately 0.27 nm, about twice the length of the covalent O—H bonds in water. The hydrogen atom is closer to the donor atom, D, to which it remains covalently bonded, than it is to the acceptor. The length of the covalent D—H bond is a bit longer than it would be if there were no hydrogen bond, because the acceptor "pulls" the hydrogen away from the donor. The strength of a hydrogen bond in water (≈5 kcal/mol) is much weaker than a covalent O—H bond (≈110 kcal/mol).

HYDROGEN BONDS AS A STABILIZING FORCE IN MACROMOLECULES

An important feature of all hydrogen bonds is directionality. In the strongest hydrogen bonds, the donor atom, the hydrogen atom, and the acceptor atom all lie in a straight line. Nonlinear hydrogen bonds are weaker than linear ones; still, multiple nonlinear hydrogen bonds help to stabilize the three-dimensional structures of many proteins. It is only because of the aggregate strength of multiple hydrogen bonds that they play a central role in the architecture of large biological molecules in aqueous solutions.

The strengths of the hydrogen bonds in proteins and nucleic acids are only 1 to 2 kcal/mol, considerably weaker than the hydrogen bonds between water molecules.. Initially, both the —OH and $-NH_2$ groups in the protein are hydrogen-bonded to water, and the formation of a hydrogen bond between these groups involves disruption of their hydrogen bonds with water. Thus the *net* change in energy in forming this —OH···N hydrogen bond will be less than the 5 kcal/mol characteristic of hydrogen bonds between water molecules.

IONIC INTERACTIONS ARE ATTRACTIONS BETWEEN OPPOSITELY CHARGED IONS

In some compounds, the bonded atoms are so different in electronegativity that the bonding electrons are never shared: these electrons are always found around the more electronegative atom. In sodium chloride (NaCl), for example, the bonding electron contributed by the sodium atom is completely transferred to the chlorine atom. Even in solid crystals of NaCl, the sodium and chlorine atoms are ionized, so it is more accurate to write the formula for the compound as Na^+Cl^-.

Because the electrons are not shared, the bonds in such compounds cannot be considered covalent. They are, rather, ionic bonds (or interactions) that result from the attraction of a positively charged ion — a cation — for a negatively charged ion — an anion. Unlike covalent or hydrogen bonds, ionic bonds do not have fixed or specific geometric orientations because the electrostatic field around an ion — its attraction for an opposite charge — is uniform in all directions. However, crystals of salts such as Na^+Cl^- do have very regular structures because that is the energetically most favorable way of packing together positive and negative ions. The force that stabilizes ionic crystals is called the *lattice energy*.

In aqueous solutions, simple ions of biological significance, such as Na^+, K^+, Ca^{2+}, Mg^{2+}, and Cl^-, do not exist as free, isolated entities. Instead, each is surrounded by a stable, tightly held shell of water molecules. An ionic interaction occurs between the ion and the oppositely charged end of the water dipole, as shown below for the K^+ ion:

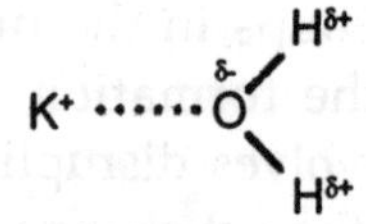

Ions play an important biological role when they pass through narrow, protein-lined pores, or channels, in membranes. For example, ionic movements through membranes are essential for the conduction of nerve impulses and for the stimulation of muscle contraction.Ions must lose their shell of water molecules in order to pass through ion channel proteins; channel proteins can then selectively admit only Na^+, or K^+, or Ca^{2+} ions, a selectivity essential for nerve function.

Most ionic compounds are quite soluble in water because a large amount of energy is released when ions tightly bind water molecules. This is known as the *energy of hydration*. Oppositely charged ions are shielded from one another by the water and tend not to recombine. Salts like Na^+Cl^- dissolve in water because the energy of hydration is greater than the lattice energy that stabilizes the crystal structure. In contrast, certain salts, such as $Ca_3(PO_4)_2$, are virtually insoluble in water; the large charges on the Ca^{2+} and $PO_4^{3''}$ ions generate a formidable lattice energy that is greater than the energy of hydration.

VAN DER WAALS INTERACTIONS ARE CAUSED BY TRANSIENT DIPOLES

When any two atoms approach each other closely, they create a weak, nonspecific attractive force that produces a van der Waals interaction, named for Dutch physicist Johannes Diderik van der Waals (1837 – 1923), who first described it. These nonspecific interactions result from the momentary random fluctuations in the distribution of the electrons of any atom, which give rise to a transient unequal distribution of electrons, that is, a transient electric dipole. If two noncovalently bonded atoms are close enough together, the transient dipole in one atom will perturb the electron cloud of the other. This perturbation generates a transient dipole in the second atom, and the two dipoles will attract each other weakly. Similarly, a polar covalent bond in one molecule will attract an oppositely oriented dipole in another.

Van der Waals interactions, involving either transient induced or permanent electric dipoles, occur in all types of molecules, both polar and nonpolar. In particular, van der Waals interactions are responsible for the cohesion between molecules of nonpolar liquids and solids, such as heptane, $CH_3-(CH_2)_5-CH_3$, that cannot form hydrogen bonds or ionic interactions with other molecules. When these stronger interactions are present, they override most of the influence of van der Waals interactions. Heptane, however, would be a gas if van der Waals interactions could not form.

The strength of van der Waals interactions decreases rapidly with increasing distance; thus these noncovalent bonds can form only when atoms are quite close to one another. However, if atoms get too close together, they become repelled by the negative charges in their outer electron shells. When the van der Waals attraction between two atoms exactly balances the repulsion between their two electron clouds, the atoms are said to be in *van der Waals contact*. Each type of atom has a van der Waals radius at which it is in van der Waals contact with other atoms. The van der Waals radius of an H atom is 0.1 nm, and the radii of O, N, C, and S atoms are between 0.14 and 0.18 nm. Two covalently bonded atoms are closer together than two atoms that are merely in van der Waals contact. For a van der Waals interaction, the internuclear distance is approximately the sum of the corresponding radii for the two participating atoms. Thus the distance between a C atom and an H atom in van der Waals contact is 0.27 nm, and between two C atoms is 0.34 nm. In general, the van der Waals radius of an atom is about twice as long as its covalent radius. For example, a C—H covalent bond is about 0.107 nm long and a C—C covalent bond is about 0.154 nm long.

The energy of the van der Waals interaction is about 1 kcal/mol, only slightly higher than the average thermal energy of molecules at 25 °C. Thus the van der Waals interaction is even weaker than the hydrogen bond, which typically has an energy of 1 – 2 kcal/mol in aqueous solutions. The attraction between two large molecules can be appreciable, however, if they have precisely complementary shapes, so that they make many van der Waals contacts when they come into proximity. Van der Waals interactions, as well as other noncovalent bonds, mediate the binding of many enzymes with their specific substrates (the substances on which an enzyme acts) and of each type of antibody with its specific antigen.

HYDROPHOBIC BONDS CAUSE NONPOLAR MOLECULES TO ADHERE TO ONE ANOTHER

Nonpolar molecules do not contain ions, possess a dipole moment, or become hydrated. Because such molecules are insoluble or almost insoluble in water, they are said to be hydrophobic (Greek, "water-fearing"). The covalent bonds between two carbon atoms and between carbon and hydrogen atoms are the most common nonpolar bonds in biological systems. Hydrocarbons — molecules made up only of carbon and hydrogen — are virtually insoluble in water. A large triacylglycerol (or triglyceride) such as tristearin, a component of animal fat, is also insoluble in water, even though its six oxygen atoms participate in some slightly polar bonds with adjacent carbon atoms. When shaken in water, tristearin forms a separate phase similar to the separation of oil from the water-based vinegar in an oil-and-vinegar salad dressing.

The force that causes hydrophobic molecules or nonpolar portions of molecules to aggregate together rather than to dissolve in water is called the hydrophobic bond. This is not a separate bonding force; rather, it is the result of the energy required to insert a nonpolar molecule into water. A nonpolar molecule cannot form hydrogen bonds with water molecules, so it distorts the usual water structure, forcing the water into a rigid cage of hydrogen-bonded molecules around it. Water molecules are normally in constant motion, and the formation of such cages restricts the motion of a number of **water** molecules; the effect is to increase the structural organization of water. This situation is energetically unfavorable because it decreases the randomness (entropy) of the population of water molecules. The role of entropy in chemical systems is discussed further in a later section.

The opposition of water molecules to having their motion restricted by forming cages around hydrophobic molecules or portions thereof is the major reason molecules such as tristearin and heptane are essentially insoluble in water and interact mainly with other hydrophobic molecules. Nonpolar molecules can also bond together, albeit weakly, through van der Waals interactions. The net result of the hydrophobic and van der Waals interactions is a very powerful tendency for hydrophobic molecules to interact with one another, and not with water.

Small hydrocarbons like butane ($CH_3-CH_2-CH_2-CH_3$) are somewhat soluble in water, because they can dissolve without disrupting the water lattice appreciably. However, 1-butanol ($CH_3-CH_2-CH_2-CH_2OH$) mixes completely with water in all proportions. The replacement of just one hydrogen atom with the polar $-OH$ group allows the molecule to form hydrogen bonds with water and greatly increases its solubility.

Simply put, *like dissolves like.* Polar molecules dissolve in polar solvents such as water, while nonpolar molecules dissolve in nonpolar solvents such as hexane.

MULTIPLE NONCOVALENT BONDS CAN CONFER BINDING SPECIFICITY

Besides contributing to the stability of large biological molecules, multiple noncovalent bonds can also confer *specificity* by determining how large molecules will fold or which regions of different molecules will bind together. All types of these weak interactions are effective only over a short range and require close contact between the reacting groups. For noncovalent bonds to form properly, there must be a complementarity between the sites on the two interacting surfaces. Almost any other arrangement of the same groups on the two surfaces would not allow the molecules to bind so tightly. Such multiple, specific interactions allow protein molecules to fold into a unique three-dimensional shape and the two chains of DNA to bind together.

PHOSPHOLIPIDS ARE AMPHIPATHIC MOLECULES

Multiple noncovalent bonds also are critical in stabilizing the structure of biomembranes, whose primary components are phospholipids. Because the essential properties of biomembranes derive from phospholipids, we first examine the chemistry of these compounds and then see how they associate into the sheetlike structures that are the foundation of biomembranes.

All phospholipids contain one or more acyl chains derived from fatty acids, which consist of a hydrocarbon chain attached to a carboxyl group (—COOH). Fatty acids are insoluble in water and salt solutions; they differ in length and in the extent and position of their double bonds. Most fatty acids have an even number of carbon atoms, usually 16, 18, or 20.

Fatty acids with no double bonds are said to be *saturated*; those with at least one double bond are *unsaturated*. Unsaturated fatty acid chains normally have one double bond, but some have two, three, or four. Two stereoisomeric configurations, cis and trans, are possible around each double bond:

A cis double bond introduces a rigid kink in the otherwise flexible straight chain of a fatty acid. In general, the fatty acids in biological systems contain only cis double bonds.

Phospholipids consist of two long-chain fatty acyl groups linked (usually by an ester bond) to small, highly hydrophilic groups. Consequently, unlike tristearin, phospholipids do not clump together in droplets but orient themselves in sheets, exposing their hydrophilic ends to the aqueous environment. Molecules in which one end (the "head") interacts with water and the other end (the "tail") is hydrophobic are said to be amphipathic (Greek, "tolerant of both"). The tendency of amphipathic molecules to form organized structures spontaneously in water is the key to the structure of cell membranes.

In phosphoglycerides, a principal class of phospholipids, fatty acyl side chains are esterified to two of the three hydroxyl groups in glycerol but the third hydroxyl group is esterified to phosphate. The simplest phospholipid, phosphatidic acid, contains only these components:

where R_1 and R_2 are fatty acyl groups.

In most phospholipids, however, the phosphate group is also esterified to a hydroxyl group on another hydrophilic compound. In phosphatidylcholine, for example, choline is attached to the phosphate. In other phosphoglycerides, the phosphate group is linked to other molecules, such as ethanolamine, the amino acid serine, or the sugar inositol. The negative charge on the phosphate as well as the charged groups or hydroxyl groups on the alcohol esterified to it interact strongly with water.

THE PHOSPHOLIPID BILAYER FORMS THE BASIC STRUCTURE OF ALL BIOMEMBRANES

When a suspension of phospholipids is mechanically dispersed in aqueous solution, they can assume three different forms: micelles, bilayer sheets, and liposomes. The type of structure formed by a pure phospholipid or a mixture of phospholipids depends on the length of the fatty acyl chains and their degree of saturation, on the temperature, on the ionic composition of the aqueous medium, and on the mode of dispersal of the phospholipids in the solution. In all three forms, hydrophobic interactions cause the fatty acyl chains to aggregate and exclude water molecules from the "core." Micelles are rarely formed from natural phosphoglycerides, whose fatty acyl chains generally are too bulky to fit into the interior of a micelle.

Under suitable conditions, phospholipids of the

composition present in cells spontaneously form symmetric sheetlike structures, called phospholipid bilayers, that are two molecules thick. Each phospholipid layer in this lamellar structure is called a *leaflet*. The hydrocarbon side chains in each leaflet minimize contact with water by aligning themselves tightly together in the centre of the bilayer, forming a hydrophobic core that is about 3 nm thick. The close packing of these hydrocarbon side chains is stabilized by van der Waals interactions between them. Ionic and hydrogen bonds stabilize the interaction of the phospholipid polar head groups with each other and with water. At neutral pH, the polar head groups in some phospholipids (e.g., phosphatidylcholine) have no net electric charge, whereas the head groups in others have a net negative charge. Nonetheless, all phospholipids can pack together into the characteristic bilayer structure.

A phospholipid bilayer can be of almost unlimited size — from micrometers (μ) to millimeters (mm) in length or width — and can contain tens of millions of phospholipid molecules. Because of their hydrophobic core, bilayers are impermeable to salts, sugars, and most other small hydrophilic molecules. Like a phospholipid bilayer, all biological membranes have a hydrophobic core, and they all separate two aqueous solutions. The plasma membrane, for example, separates the interior of the cell from its surroundings. Similarly, the membranes that surround the organelles of eukaryotic cells separate one aqueous phase — the cell cytosol — from another — the interior of the organelle. Several types of evidence indicate that the phospholipid bilayer is the basic structural unit of nearly all biomembranes. Associated with membrane phospholipids are various proteins that help confer unique properties on each type of membrane.

CHEMICAL EQUILIBRIUM

We now shift our discussion to chemical reactions in which bonds, primarily covalent bonds, are broken and re-formed. At any one time several hundred different kinds of chemical reactions are occurring simultaneously in every cell, and any chemical can, in principle, undergo multiple chemical reactions. Both the *extent* to which a reaction can proceed and the *rate* at which it actually takes place determine which reactions actually occur in cells.

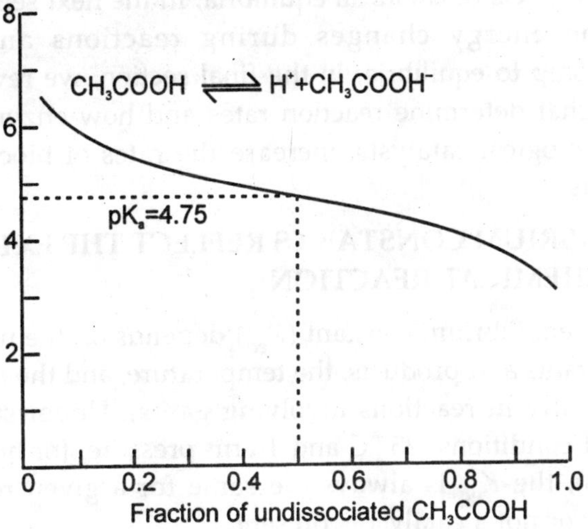

Fig. The titration curve of acetic acid (CH_3COOH).

When reactants first come together — before any products have been formed — their rate of reaction is determined in part by their initial concentrations. As the reaction products accumulate the concentration of each reactant decreases and so does the reaction rate. Meanwhile, some of the product molecules begin to participate in the reverse reaction, which re-forms the reactants. This reaction is slow at first but speeds up as the concentration of products increases. Eventually, the rates of the forward and reverse reactions become equal, so that the concentrations of reactants and products stop changing. The mixture is then said to be in chemical equilibrium.

At equilibrium the ratio of products to reactants, called the equilibrium constant, is a fixed value that is independent of the rate at which the reaction occurs. The rate of a chemical reaction can be increased by a catalyst, a substance that brings reactants together and accelerates their interactions but is not permanently changed during a reaction. This function is aptly reflected in the Chinese term for catalyst, *tsoo mei,* which literally means "marriage broker." In this section, we discuss several aspects of chemical equilibria; in the next section, we examine energy changes during reactions and their relationship to equilibria. In the final section, we review the factors that determine reaction rates and how enzymes, the cell's biological catalysts, increase the rates of biochemical reactions.

EQUILIBRIUM CONSTANTS REFLECT THE EXTENT OF A CHEMICAL REACTION

The equilibrium constant (K_{eq}) depends on the nature of the reactants and products, the temperature, and the pressure (particularly in reactions involving gases). Under standard physical conditions (25 °C and 1 atm pressure, for biological systems), the K_{eq} is always the same for a given reaction, whether or not a catalyst is present.

For the simple reaction $A + B \rightleftharpoons X + Y$, the equilibrium constant is given by

$$K_{eq} = \frac{[X][Y]}{[A][B]} \qquad (2-1)$$

Where brackets indicate equilibrium concentrations. In general, for a reaction

$$aA + bB + cC + \cdots \rightleftharpoons zZ + yY + xX + \cdots \qquad (2-2)$$

Where capital letters represent particular molecules or atoms and lowercase letters represent the number of each in the reaction formula, the equilibrium constant is given by

$$K_{eq} = \frac{[X]^x [Y]^y [Z]^z}{[A]^a [B]^b [C]^c} \qquad (2-3)$$

In the generalized reaction (2-2) above, the rate of the forward (left to right) reaction will be

$$Rate_{forward} = k_f [A]^a [B]^b [C]^c$$

Where k_f is the rate constant for the forward reaction. Similarly, the rate of the reverse (right to left) reaction is

$$Rate_{reverse} = k_r [X]^x [Y]^y [Z]^z$$

Where k_r is the rate constant for the reverse reaction. Since at equilibrium the rate of the forward reaction equals that of the reverse reaction, we can write

$$k_f [A]^a [B]^b [C]^c = k_r [X]^x [Y]^y [Z]^z$$

Which can be rearranged to give

$$\frac{k_f}{k_r} = \frac{[X]^x [Y]^y [Z]^z}{[A]^a [B]^b [C]^c}$$

Comparing this expression with Equation 2-3, we see the important relationship

$$k_{eq} = \frac{k_f}{k_r} \qquad (2-4)$$

In other words, the equilibrium constant equals the ratio of the forward and reverse rate constants. Because a catalyst accelerates the rates of the forward and reverse reactions by the same factor, it does not change the value of k_f/k_r. Thus, as noted above, catalysts do not alter the equilibrium constant, which depends only on the chemical properties of the molecules involved and on the temperature and pressure.

To illustrate several points concerning equilibrium, we shall use a fairly simple biochemical reaction: the interconversion of the compounds glyceraldehyde 3-phosphate (*G3P*) and dihydroxyacetone phosphate (*DHAP*). This reaction, which occurs during the breakdown of glucose, is catalyzed by the enzyme triosephosphate isomerase:

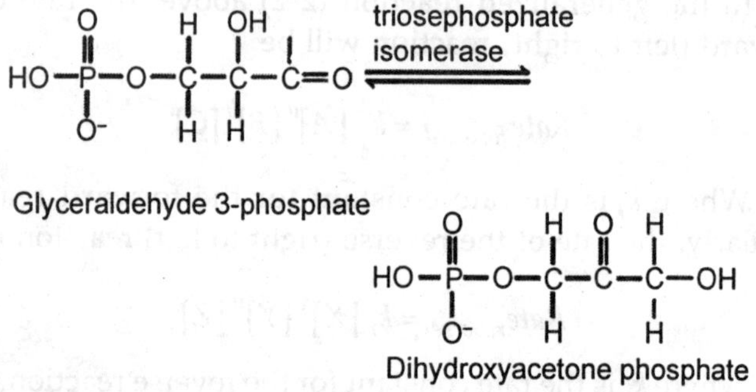

The equilibrium constant for this reaction under standard conditions is

$$k_{eq} = \frac{[DHAP]}{[G3P]} = 22.2$$

Thus the ratio of the concentrations of *G3P* and *DHAP* is 1:22.2 when the reaction reaches equilibrium. In practice, one measures the concentrations of reactants and products after a reaction has reached equilibrium, and uses these values to calculate the equilibrium constant.

In the presence of an enzyme or other catalyst, the reaction rate may increase, but the final ratio of product to reactant will always be the same. The magnitude of the equilibrium constant has no bearing on the rate of the reaction or on whether the reaction will take place at all under normal conditions. Despite the large equilibrium constant for the conversion of *G3P* to *DHAP*, for example, so much energy is required to rearrange the bonds that no detectable reaction actually occurs in an aqueous solution in the absence of an enzyme or other catalyst.

When a reaction involves a *single* reactant and a *single* product, the ratio of the product concentration to reactant concentration at equilibrium is equal to the equilibrium constant K_{eq} and is independent of the initial concentrations. When a reaction involves *multiple* reactants and/or products, the *equilibrium concentration* of any one product or reactant depends on the initial concentrations of all reactants and products as well as on the equilibrium constant. Consider, for example, the hydrolysis (cleavage by addition of water) of the dipeptide glycylalanine (*GA*) to glycine (*G*) and alanine (*A*):

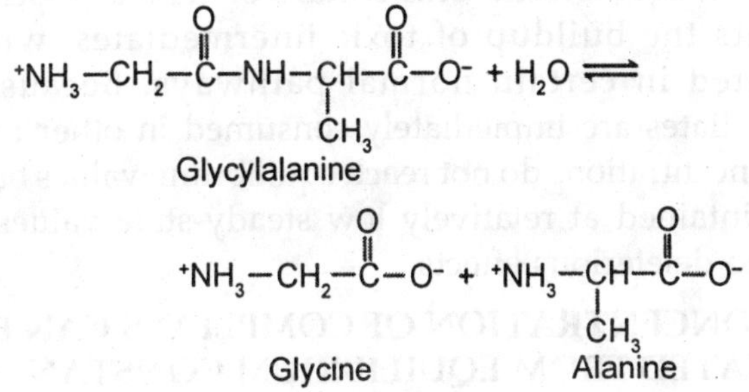

Here,

$$k_{eq} = \frac{[G][A]}{[GA]}$$

(The concentration of water does not change significantly during normal aqueous chemical reactions and, by convention, is not included in the calculation of equilibrium ratios.) The equilibrium is strongly in the direction of the formation of glycine and alanine. In other words, most of the glycylalanine is hydrolyzed at equilibrium. However, an excess of one of the products can drive the reaction in the reverse direction. For instance, suppose that the initial reaction mixture contains a small amount of glycylalanine and a large amount of alanine. As the reaction proceeds, the total concentration of alanine [*A*] will always greatly exceed the concentration of glycine [*G*] produced by hydrolysis. This must reduce the equilibrium ratio of [*G*] to [*GA*] because K_{eq} remains constant. Thus, the reaction $GA \rightleftharpoons G + A$ can be driven to the left by an excess of the product alanine. More generally, in reactions involving more than one reactant or product, changes in the concentration of any one reactant or product will affect the concentrations at equilibrium of all the reactants and products.

Under appropriate conditions, individual biochemical reactions carried out in a test tube eventually will reach equilibrium. Within cells, however, many reactions are linked in pathways in which a product of one reaction is frequently used quickly as a substrate for another. In this more complex situation, reactants and products generally are in *steady state* but not in equilibrium. This feature of cellular biochemistry prevents the buildup of toxic intermediates, which are generated in certain normal pathways. Because these intermediates are immediately consumed in other reactions, their concentrations do not reach equilibrium values but rather are maintained at relatively low steady-state values, which cause no deleterious effects.

THE CONCENTRATION OF COMPLEXES CAN BE ESTIMATED FROM EQUILIBRIUM CONSTANTS FOR BINDING REACTIONS

The reactions discussed above exemplify the formation or cleavage of covalent bonds. Many important reactions, however, involve the binding of one molecule to another, mediated by various noncovalent interactions. A common example is the binding of a ligand (e.g., the hormone insulin or adrenaline) to its receptor on the surface of a cell, triggering a biological response. Another example is the binding of a protein to a specific sequence of base pairs in a molecule of DNA, which frequently causes the expression of a nearby gene to turn on or off. Such DNA-binding proteins are discussed in detail in later. Here we focus on how the equilibrium constant for binding of a protein to DNA can be used to calculate the extent to which the protein is bound to DNA in a cell.

For the binding reaction $P + D \rightleftharpoons PD$, where *PD* is the specific complex of a protein (*P*) and DNA (*D*), the equilibrium constant is customarily defined as the ratio of reactants to products; it is also called the *dissociation constant* and is given by

$$k_{eq} = \frac{[P][D]}{[PD]} \qquad (2-5)$$

the inverse of the reactions in the previous section. This means that when the concentration of protein, [*P*], equals K_{eq}, then half the DNA will contain a bound protein and half will not; that is, $[D] = [PD]$. Typical reactions in which a protein binds to a specific DNA sequence have a K_{eq} of 10^{-10} *M*, where M symbolizes *molarity*, or moles per liter (*mol/L*). We can relate the magnitude of this equilibrium constant to events within a bacterial cell, which contains *one molecule* of DNA. Assume, for simplicity, that the cell also contains *ten molecules* of the DNA-binding protein *P*. Further assume that the cell is a cylinder 2 μm long and 1 μm in diameter. Thus its volume will be 1.5×10^{-15} *L*, and the total concentration of the one molecule of DNA (*D*) within the cell, will be

$$\frac{(1\,molecule / 1.5 \times 10^{-15}\,L)}{(6.02 \times 10^{23}\,molecules / mol)} = 1.1 \times 10^{-9}\,M$$

Similarly, the total concentration of the 10 molecules of *P*, the protein, will be 1.1×10^{-8} *M*.

If we define $[D]$ to be the concentration of free DNA, unbound to protein, and $[PD]$ to be the concentration of the DNA-protein complex, then the total concentration of DNA is $[D] + [PD]$. Similarly, if $[P]$ is defined as the concentration of free protein, unbound to DNA, then the total concentration of protein is $[P] + [PD]$. By transposing terms and using the total concentrations calculated above, we can write the following equations:

$[D] = 1.1 \times 10^{-9}\,M - [PD]$ and

$[P] = 1.1 \times 10^{-8}\,M - [PD]$ and

Substituting these expressions into Equation 2-5 and rearranging gives

$$[PD] = \frac{[P][D]}{K_{eq}} = \frac{\left(1.1 \times 10^{-8}\,M - [PD]\right)\left(1.1 \times 10^{-9}\,M - [PD]\right)}{10^{-10}\,M}$$

Solving this quadratic equation for $[PD]$ gives 1.089×10^{-9} M, the concentration of the protein-DNA complex at equilibrium. Substituting this value into the above equations for $[P]$, the concentration of free (unbound) protein, and $[D]$, the concentration of unbound DNA, at equilibrium, we get

$[P] = 1.1 \times 10^{-8}\,M - 1.089 \times 10^{-9}\,M = 9.911 \times 10^{-9}\,M$

$[P] = 1.1\times10^{-8}\,M - 1.089\times10^{-9}\,M = 1.1\times10^{-11}\,M$

From these values, we can now calculate the ratio of bound to total DNA:

$$\frac{[PD]}{[PD]+[D]} = \frac{1.089\times10^{-9}\,M}{1.1\times10^{-9}\,M} = 0.99$$

Thus 99 percent of the time this specific sequence of DNA will have a molecule of protein bound to it, and 1 percent of the time it will not, even though the cell contains only 10 molecules of the protein! This example illustrates the general approach for predicting the abundance of intracellular complexes if one knows the equilibrium constants for the binding reactions and the numbers of the participating molecules within a cell.

BIOLOGICAL FLUIDS HAVE CHARACTERISTIC PH VALUES

The solvent inside cells and in all extracellular fluids is water. An important characteristic of any aqueous solution is the concentration of positively charged hydrogen ions (H^+) and negatively charged hydroxyl ions (OH^-). Because these ions are the dissociation products of H_2O, they are constituents of all living systems, and they are liberated by many reactions that take place between organic molecules within cells.

When a water molecule dissociates, one of its polar $H{-}O$ bonds breaks. The resulting hydrogen ion, often referred to as a *proton*, has a short lifetime as a free particle and quickly combines with a water molecule to form a hydronium ion (H_3O^+). For convenience however, we refer to the concentration of hydrogen ions in a solution, $[H^+]$, even though we really mean the concentration of hydronium ions, $[H_3O^+]$. The dissociation of water is a reversible reaction,

$$H_2O \rightleftharpoons H^+ + OH^-$$

and at 25 °C,

$$\left[H^+\right]\left[OH^-\right] = 10^{-14}\,M^2$$

In pure water, $[H^+]$ = [O≈] = 10"7 M.

The concentration of hydrogen ions in a solution is expressed conventionally as its *pH*:

$$pH = -\log\left[H^+\right] = \log\frac{1}{\left[H^+\right]}$$

In pure water at 25 °C, $[H^+] = 10^{-7}$ M, so

$$pH = -\log 10^{-7} = 7.0$$

On the *pH* scale, 7.0 is considered neutral: *pH* values below 7.0 indicate acidic solutions and values above 7.0 indicate basic (alkaline) solutions. In a 0.1 M solution of hydrogen chloride (*HCl*) in water, $[H^+]$ = 0.1 *M* because virtually all the *HCl* has dissociated into H^+ and Cl^- ions. For this solution pH = –log 0.1 = 1.0. In fact, pH values can be less than zero, since a 10 *M* solution of *HCl* will have a *pH* of "1.

One of the most important properties of a biological fluid is its *pH*. The cytosol of cells normally has a *pH* of about 7.2. In certain organelles of eukaryotic cells, such as the lysosomes and vacuoles, the *pH* is much lower, about 5; this corresponds to a H^+ concentration more than 100 times higher than that in the cytosol. Lysosomes contain many degradative enzymes that function optimally in an acidic environment, whereas their action is inhibited in the near-neutral environment of the cytosol. Maintenance of a specific *pH* is imperative for some cellular structures to function properly. On the other hand, dramatic shifts in cellular *pH* may play an important role in controlling cellular activity. For example, the pH of the cytosol of an unfertilized sea urchin egg is 6.6. Within 1 minute of fertilization, however, the *pH* rises to 7.2; that is, the H^+ concentration decreases to about one-fourth its original value. The change in *pH* is necessary for subsequent growth and division of the egg.

HYDROGEN IONS ARE RELEASED BY ACIDS AND TAKEN UP BY BASES

In general, any molecule or ion that tends to release a hydrogen ion is called an acid, and any molecule or ion that readily combines with a hydrogen ion is called a base. Thus hydrogen chloride is an acid. The hydroxyl ion is a base, as is ammonia (NH_3), which readily picks up a hydrogen ion to become an ammonium ion (NH_4^+). Many organic molecules are acidic because they have a carboxyl group (*—COOH*), which tends to dissociate to form the negatively charged carboxylate ion (*—COO*$^-$):

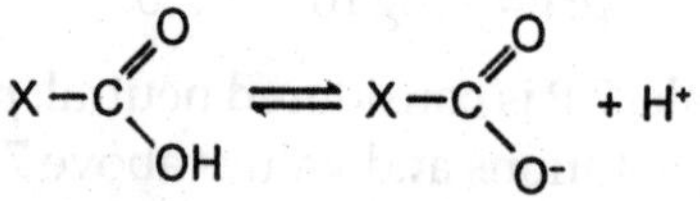

where X represents the rest of the molecule. The amino group (*—NH*$_2$), a part of many important biological molecules, is a base because, like ammonia, it can take up a hydrogen ion:

$$X-NH_2+H^+ \rightleftharpoons X-NH_3^+$$

When acid is added to a solution, $[H^+]$ increases (the *pH* goes down). Consequently, $[OH^-]$ decreases because hydroxyl ions readily combine with the hydrogen ions to form water. Conversely, when a base is added to a solution, $[H^+]$ decreases (the pH goes up). Because $[H^+][OH^-] = 10^{-14}\ M^2$, any increase in $[H^+]$ is coupled with a decrease in $[OH^-]$, and vice versa. No matter how acidic or alkaline a solution is, it always contains both ions: neither [O≈] nor $[H^+]$ is ever zero. For example, if $[H^+] = 0.1\ M$ ($pH = 1.0$), then $[OH^-] = 10$"^{13} *M*.

The degree to which a dissolved acid releases hydrogen ions or a base takes them up depends partly on the *pH* of the solution. Amino acids provide an example of this phenomenon. These molecules have the general formula

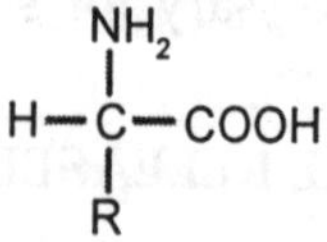

where R represents the rest of the molecule. In neutral solutions ($pH = 7.0$), amino acids exist predominantly in the doubly ionized form

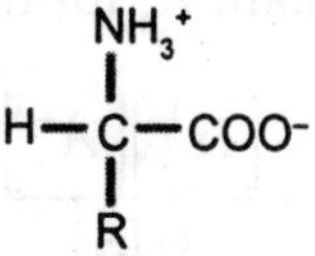

Such a molecule, containing both a positive and a negative ion, is called a *zwitterion*. Zwitterions, having no net charge, are neutral.

In solutions at low *pH*, carboxylate ions ($-COO^-$) recombine with the abundant hydrogen ions, so that the predominant form of the amino acid molecule is

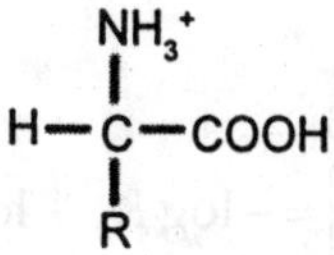

At high *pH*, the scarcity of hydrogen ions decreases the chance that an amino group or a carboxylate ion will pick up a hydrogen ion, so that the predominant form of an amino acid molecule is

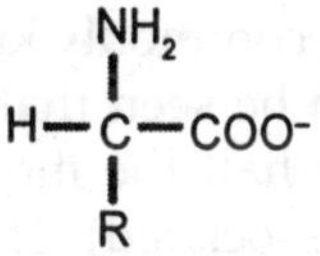

The Henderson-Hasselbalch Equation Relates pH and K_{eq} of an Acid-Base System

Many molecules used by cells have multiple acidic or basic groups, each of which can release or take up a proton. In the laboratory, it is often essential to know the precise state of dissociation of each of these groups at various *pH* values. The dissociation of an acid group *HA*, such as acetic acid (CH_3COOH), is described by

$$HA \rightleftharpoons H^+ + A^-$$

The equilibrium constant K_a for this reaction is

$$K_a = \frac{\left[H^+\right]\left[A^-\right]}{[HA]}$$

By taking the logarithm of both sides and rearranging the result, we can derive a very useful relation between the equilibrium constant and *pH* as follows:

$$\log K_a = \log\frac{\left[H^+\right]\left[A^-\right]}{[HA]} = \log\left[H^+\right] + \log\frac{\left[A^-\right]}{[HA]}$$

or

$$-\log\left[H^+\right] = -\log K_a + \log\frac{\left[A^-\right]}{[HA]}$$

Substituting *pH* for $-\log [H^+]$ and pK_a for $-\log K_a$, we have

$$pH = pK_a + \log\frac{\left[A^-\right]}{[HA]} \qquad (2-6)$$

From this expression, commonly known as the *Henderson-Hasselbalch equation,* it can be seen that the pK_a of any acid is equal to the *pH* at which half the molecules are dissociated and half are neutral (undissociated). This is because when $pK_a = pH$, then $\log ([A^-]/[HA]) = 0$, and therefore $[A^-] = [HA]$. The Henderson-Hasselbalch equation allows us to calculate the degree of dissociation of an acid if both the *pH* of the solution and the pK_a of the acid are known. Experimentally, by measuring the concentration of A^- and of *HA* as a function of the solution's *pH,* one can calculate the pK_a of the acid and thus the equilibrium constant for the dissociation reaction.

BUFFERS MAINTAIN THE PH OF INTRACELLULAR AND EXTRACELLULAR FLUIDS

A growing cell must maintain a constant pH in the cytoplasm of about 7.2 – 7.4 despite the production, by metabolism, of many acids, such as lactic acid and CO_2, which reacts with water to form carbonic acid (H_2CO_3). Cells have a reservoir of weak bases and weak acids, called buffers, which ensure that the cell's pH remains relatively constant. Buffers do this by "soaking up" H^+ or OH^- when these ions are added to the cell or are produced by metabolism.

If additional acid (or base) is added to a solution of an acid (or a base) at its pK_a value (a 1:1 mixture of HA and A^-), the pH of the solution changes, but it changes less than it would if the original acid (or base) had not been present. This is because protons released by the added acid are taken up by the original A^- form of the acid; likewise, hydroxyl ions generated by the added base are neutralized by protons released by the original HA.

This ability of a buffer to minimize changes in pH, its *buffering capacity*, depends on the relationship between its pK_a value and the pH. To understand this point, we need to recognize the effect of pH on the fraction of molecules in the undissociated form (HA). The *titration curve* for acetic acid at one pH unit below the pK_a of an acid, 91 percent of the molecules are in the HA form; at one pH unit above the pK_a, 91 percent are in the A^- form. Thus the buffering capacity of weak acids and bases declines rapidly at more than one pH unit from their pK_a values. In other words, the addition of the same number of moles of acid to a solution containing a mixture of HA and A^- that is at a pH near the pK_a of the acid will cause less of a pH change than it would if the HA and A^- were not present or if the pH were far from the pK_a value.

All biological systems contain one or more buffers. Phosphoric acid (H_3PO_4) is a physiologically important buffer; phosphate ions are present in considerable quantities in cells and are an important factor in maintaining, or buffering, the pH of the cytosol. Phosphoric acid

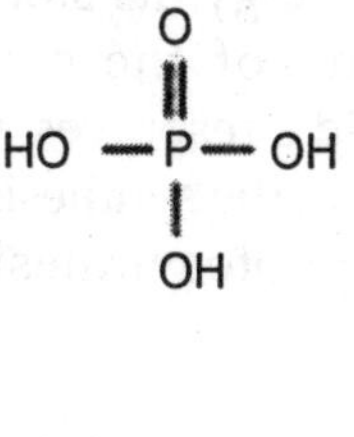

has three groups that are capable of dissociating, but the three protons do not dissociate simultaneously. Loss of each proton can be described by a discrete dissociation reaction and pK_a. The titration curve for phosphoric acid shows that the pK_a for the dissociation of the second proton is pH 7.2, similar to the pH of the cytosol. Because $pK_a = 7.2$ for the reaction $H_2PO_4^- \rightleftharpoons HPO_4^{2-} + H^+$, at pH 7.2 about 50 percent of cellular phosphate is $H_2PO_4^-$ and 50 percent is HPO_4^{2-} according to the Henderson-Hasselbalch equation. (The actual proportions, calculated by exact solution of the Henderson-Hasselbalch equation using all three dissociation constants are 0.499973 as $H_2PO_4^-$, 0.499973 as HPO_4^{2-}, 0.0000039 as H_3PO_4, and 0.0000016 as PO_4^{3-}.) Thus, phosphate is an excellent buffer at pH values around 7.2, the approximate pH of the cytosol of cells, and at pH 7.4, the pH of human blood.

In nucleic acids, phosphate is found as a diester linked to two carbon atoms of adjacent ribose sugars:

The pK_a for the dissociation of the single $-OH$ proton is about 3, which is similar to the pK_a for the dissociation of the first proton from phosphoric acid. Therefore, each phosphate residue in deoxyribonucleic acid (DNA) or ribonucleic acid (RNA) is dissociated and carries a negative charge at neutral pH, which is why DNA and RNA are called nucleic *acids:*

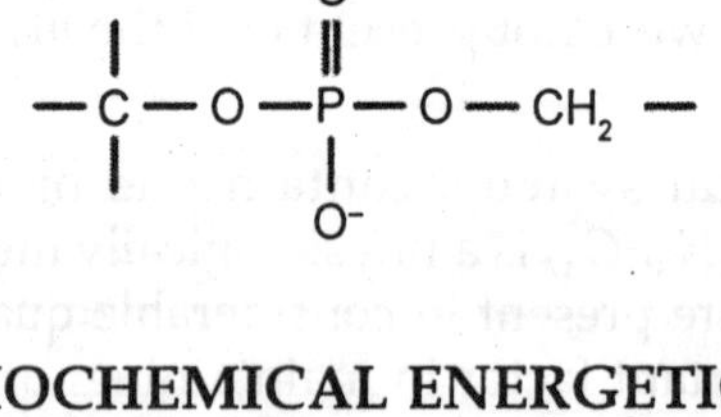

BIOCHEMICAL ENERGETICS

The production of energy, its storage, and its use are as central to the economy of the cell as they are to the management of the world's resources. Cells require energy to do all their work, including the synthesis of sugars from carbon dioxide and water in photosynthesis, the contraction of muscles, and the replication of DNA. Energy may be defined as the ability to do work, a concept that is easy to grasp when it is applied to automobile engines and electric power plants. When we consider the energy associated with chemical bonds and chemical reactions within cells, however, the concept of work becomes less intuitive.

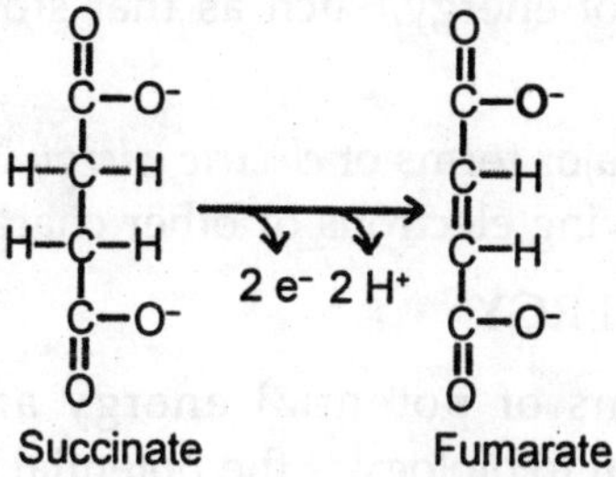

Fig. Succinate is converted to fumarate by the loss of two electrons and two protons.

Living Systems Use Various Forms of Energy, Which Are Interconvertible

There are two principal forms of energy: kinetic and potential. *Kinetic energy* is the energy of movement — the motion of molecules, for example. The second form of energy, *potential energy,* or stored energy, is more important in the study of biological or chemical systems.

KINETIC ENERGY

Heat, or *thermal energy,* is a form of kinetic energy — the energy of the motion of molecules. For heat to do work, it must flow from a region of higher temperature — where the average speed of molecular motion is greater — to one of lower temperature. Differences in temperature often exist between the internal and external environments of cells; however, cells generally cannot harness these heat differentials to do work. Even in warm-blooded animals that have evolved a mechanism for thermoregulation, the kinetic energy of molecules is used chiefly to maintain constant organismic temperatures.

Radiant energy is the kinetic energy of photons, or waves of light, and is critical to biology. Radiant energy can be converted to thermal energy, for instance when light is absorbed by molecules and the energy is converted to molecular motion. In the process of photosynthesis, light energy is absorbed by chlorophyll and is ultimately converted into other types of energy, such as that stored in covalent chemical bonds.

One of the major forms of *electric energy* is also kinetic — the energy of moving electrons or other charged particles.

POTENTIAL ENERGY

Several forms of potential energy are biologically significant. Central to biology is the potential energy *stored in the bonds* connecting atoms in molecules. Indeed, most of the biochemical reactions described in this book involve the making or breaking of at least one covalent chemical bond. We recognize this energy when chemicals undergo energy-releasing reactions. The sugar glucose, for example, is high in potential energy. Cells degrade glucose continuously, and the energy released when glucose is metabolized is harnessed to do many kinds of work.

A second biologically important form of potential energy, to which we shall refer often, is the energy in a *concentration gradient*. When the concentration of a substance on one side of a permeable barrier, such as a membrane, is different from that on the other side, the result is a concentration gradient. All cells form concentration gradients between their interior and the external fluids by selectively exchanging nutrients, waste products, and ions with their surroundings. Also, compartments within cells frequently contain different concentrations of ions and other molecules; the concentration of protons within a lysosome, as we saw in the last section, is about 500 times that of the cytosol.

A third form of potential energy in cells is an *electric potential* — the energy of charge separation. For instance, there is a gradient of electric charge of ≈200,000 volts per cm across the outer, or "plasma," membrane of virtually all cells.

INTERCONVERTIBILITY OF ALL FORMS OF ENERGY

According to the first law of thermodynamics, energy is neither created nor destroyed, but can be converted from one form to another.* In photosynthesis, for example, as we have just seen, the radiant energy of light is transformed into the chemical potential energy of the covalent bonds between the atoms in a sucrose or starch molecule. In muscles and nerves, chemical potential energy stored in covalent bonds is transformed, respectively, into kinetic and electric energy. In all cells, chemical potential energy, released by breakage of certain chemical bonds, is used to generate potential energy in the form of concentration and electric potential gradients. Similarly, energy stored in chemical concentration gradients or electric potential gradients is used to synthesize chemical bonds, or to transport other molecules "uphill" against a concentration gradient. This latter process occurs during the transport of nutrients such as glucose into certain cells and transport of many waste products out of cells. Because all forms of energy are interconvertible, they can be expressed in the same units of measurement, such as the calorie or kilocalorie.

THE CHANGE IN FREE ENERGY ΔG DETERMINES THE DIRECTION OF A CHEMICAL REACTION

Because biological systems are generally held at constant temperature and pressure, it is possible to predict the direction of a chemical reaction by using a measure of potential energy called free energy, or *G*, after the great American chemist Josiah Willard Gibbs (1839 – 1903), a founder of the science of thermodynamics. Gibbs showed that under conditions of constant pressure and temperature, as generally found in biological systems, "all systems change in such a way that free energy is minimized." In general, we are interested in what happens to the free energy when one molecule or molecular configuration is changed into another. Thus our concern is with relative, rather than absolute, values of free energy — in particular, with the difference between the values before and after the change. This free-energy change ΔG, where Δ stands for difference, is given by

$$\Delta G = G_{products} - G_{reac\tan ts}$$

In mathematical terms, Gibbs's law — that systems change to minimize free energy — is a set of statements about ΔG:

- If ΔG is negative for a chemical reaction or mechanical process, the forward reaction or process (from left to right as written) will tend to occur spontaneously.
- If ΔG is positive, the reverse reaction (from right to left as written) will tend to occur.
- If ΔG is zero, both forward and reverse reactions occur at equal rates; the reaction is at equilibrium.

The value of ΔG, like the equilibrium constant, is independent of the reaction mechanism and rate. Reactions with negative ΔG values that have very slow rate constants may not occur, for practical purposes, unless a catalyst is present, but the presence of a catalyst does not affect the value of ΔG.

The ΔG of a Reaction Depends on Changes in Enthalpy (Bond Energy) and Entropy

At any constant temperature and pressure, two factors determine the ΔG of a reaction and thus whether the reaction will tend to occur: the change in bond energy between reactants and products and the change in the randomness of the system. Gibbs showed that free energy can be defined as

$$G = H - TS$$

where H is the bond energy, or enthalpy, of the system; T is its temperature in degrees Kelvin (K); and S is a measure of randomness, called entropy. If temperature remains constant, a reaction proceeds spontaneously only if the freeenergy change ΔG in the following equation is negative:

$$\Delta G = \Delta H - T\Delta S \qquad (2\text{-}7)$$

The enthalpy H of reactants or of products is equal to their total bond energies; the overall change in enthalpy ΔH is equal to the overall change in bond energies. In an exothermic reaction, the products contain less bond energy than the reactants, the liberated energy is usually converted to heat (the energy of molecular motion), and ΔH is negative. In an endothermic reaction, the products contain more bond energy than the reactants, heat is absorbed, and ΔH is positive. Reactions tend to proceed if they liberate energy (if $\Delta H < 0$), but this is only one of two important parameters of free energy to consider; the other is entropy.

Entropy S is a measure of the degree of randomness or disorder of a system. Entropy increases as a system becomes more disordered and decreases as it becomes more structured. Consider, for example, the diffusion of solutes from one solution into another one in which their concentration is lower. This important biological reaction is driven only by an increase in entropy; in such a process ΔH is near zero. To see this, suppose that a 0.1 M solution of glucose is separated from a large volume of water by a membrane through which glucose can diffuse. Diffusion of glucose molecules across the membrane will give them more room in which to move, with the result that the randomness, or entropy, of the system is increased. Maximum entropy is achieved when all molecules can diffuse freely over the largest possible volume — that is, when the concentration of glucose molecules is the same on both sides of the membrane. If the degree of hydration of glucose does not change significantly on dilution, ΔH will be approximately zero; the negative free energy of the reaction in which glucose molecules are liberated to diffuse over a larger volume will be due solely to the positive value of ΔS in Equation 2-7.

As mentioned previously, the formation of hydrophobic bonds is driven primarily by a change in entropy. That is, if a long hydrophobic molecule, such as heptane or tristearin, is dissolved in water, the water molecules are forced to form a cage around it, restricting their free motion. This imposes a high degree of order on their arrangement and lowers the entropy of the system ($\Delta S < 0$). Because the entropy change is negative, hydrophobic molecules do not dissolve well in aqueous solutions and tend to stay associated with one another.

We can summarize the relationships between free energy, enthalpy, and entropy as follows:

- An exothermic reaction ($\Delta H < 0$) that increases entropy ($\Delta S > 0$) occurs spontaneously ($\Delta G < 0$).
- An endothermic reaction ($\Delta H > 0$) will occur spontaneously if ΔS increases enough so that the $T\ \Delta S$ term can overcome the positive ΔH.
- If the conversion of reactants into products results in no change in free energy ($\Delta G = 0$), then the system is at equilibrium; that is, any conversion of reactants to products is balanced by an equal conversion of products to reactants.

Many biological reactions lead to an increase in order, and thus a decrease in entropy ($\Delta S < 0$). An obvious example is the reaction that links amino acids together to form a protein. A solution of protein molecules has a lower entropy than does a solution of the same amino acids unlinked, because the free movement of any amino acid in a protein is restricted when it is bound in a long chain. For the linking reaction to proceed, a compensatory decrease in free energy must occur elsewhere in the system.

SEVERAL PARAMETERS AFFECT THE ΔG OF A REACTION

The change in free energy of a reaction (ΔG) is influenced by temperature, pressure, and the initial concentrations of reactants and products. Most biological reactions — like others that take place in aqueous solutions — also are affected by the pH of the solution.

The standard free-energy change of a reaction $\Delta G^{\circ\prime}$ is the value of the change in free energy under the conditions of 298 *K* (25 °C), 1 atm pressure, *pH* 7.0 (as in pure water), and initial concentrations of 1 M for all reactants and products except protons, which are kept at *pH* 7.0. Values of $\Delta G°'$ for some typical biochemical reactions. The sign of $\Delta G°'$ depends on the direction in which the reaction is written. If the reaction A→ B has a $\Delta G°'$ of "*x kcal/mol,* then the reverse reaction B→A will have a $\Delta G°'$ value of +*x* kcal/mol.

Most biological reactions differ from standard conditions, particularly in the concentrations of reactants. However, we can estimate free-energy changes for different temperatures and initial concentrations, using the equation

$$\Delta G = \Delta G°' + RT\, InQ = \Delta G°' + RT\, in \frac{[products]}{[reac\tan ts]}$$

where R is the gas constant of 1.987 cal/(degree · mol), T is the temperature (in degrees Kelvin), and Q is the initial ratio of products to reactants, which is expressed as in Equation 2-1 defining the equilibrium constant. Again using as our example the interconversion of glyceraldehyde 3-phosphate (G3P) and dihydroxyacetone phosphate (DHAP)

$$G3P \rightleftharpoons DHAP$$

we have Q = [DHAP]/[G3P] and $\Delta G°'$ = "1840 cal/mol. Equation 2-8 for ΔG then becomes

$$\Delta G = -1840 + 1.987\, T\, In \frac{[DHAP]}{[G3P]}$$

from which we can calculate ΔG for any set of concentrations of DHAP and G3P. If the initial concentrations of both DHAP and G3P are 1 M, then $\Delta G = \Delta G°'$ = "1840 cal/mol, because RT ln 1 = 0. The reaction will tend to proceed from left to right, in the direction of formation of DHAP. If, however, the initial concentration of DHAP is 0.1 M and that of G3P is 0.001 M, with other conditions being standard, then Q = 0.1/0.001 = 100, and

$$\Delta G = -1840 + (1.987)(298)\, In(100) = +887\ cal/mol$$

Clearly, the reaction will now proceed in the direction of formation of G3P.

In a reaction $A + B \rightleftharpoons C$, in which two molecules combine to form a third, the equation for ΔG becomes

$$\Delta G = \Delta G^{\circ\prime} + RT\, In Q = \Delta G^{\circ\prime} + RT\, in \frac{[C]}{[A][B]}$$

The direction of the reaction will shift more toward the right (toward formation of *C)* if either *[A]* or *[B]* is increased.

The ΔG°′ of a Reaction Can Be Calculated from Its *Keq*

A chemical mixture at equilibrium is already in a state of minimal free energy: no free energy is being generated or released. Thus, for a system at equilibrium, we can write

$$0 = \Delta G = \Delta G^{\circ\prime} + RT\, In Q$$

At equilibrium the value of *Q* is the equilibrium constant K_{eq}, so that

$$\Delta G^{\circ\prime} = -RT\, In\, K_{eq}$$

Expressed in terms of base 10 logarithms, this equation becomes

$$\Delta G^{\circ\prime} = -2.3RT \log K_{eq}$$

or

$$\Delta G^{\circ\prime} = -2.3(1.987)(298) \log K_{eq}$$

$$= -1362 \log Keq \qquad (2-9)$$

under standard conditions. Thus, if the concentrations of reactants and products at equilibrium (i.e., the K_{eq}) are determined, the value of ΔG°′ can be calculated. For example, we saw earlier that K_{eq} equals 22.2 for the interconversion of glyceraldehyde 3-phosphate to dihydroxyacetone phosphate (G3P $_y{}^z$ DHAP) under standard conditions. Substituting this value into Equation 2-9, we can easily calculate the ΔG°′ for this reaction as –1840 cal/mol.

By rearranging Equation 2-9 and taking the antilogarithm, we obtain

$$K_{eq} = 10^{-(\Delta G°'/2.3RT)} \quad (2-10)$$

From this expression, it is clear that if $\Delta G°'$ is negative, then the exponent will be positive and hence K_{eq} will be greater than 1; that is, the formation of products from reactants is favored. Conversely, if $\Delta G°'$ is positive, then the exponent will be negative and K_{eq} will be less than 1.

Although a chemical equilibrium appears to be unchanging and static, it is actually a dynamic state. The forward and the reverse reactions proceed at exactly the same rate, thereby canceling each other out. As noted earlier, when an enzyme or some other catalyst speeds up a reaction, it also speeds up the reverse reaction; thus equilibrium is reached sooner than it is when the reaction is not catalyzed. However, the equilibrium constant and $\Delta°'$ of a reaction are the *same* in the presence and absence of a catalyst.

CELLS MUST EXPEND ENERGY TO GENERATE CONCENTRATION GRADIENTS

A cell must often accumulate chemicals, such as glucose and K^+ ions, in greater concentrations than exist in its environment. Consequently, the cell must transport these chemicals against a concentration gradient. To find the amount of energy required to transfer 1 mole of a substance from outside the cell to inside the cell, we use Equation 2-8 relating ΔG to the concentration of reactants and products. Because this simple transport reaction does not involve making or breaking covalent bonds and no heat is taken up or released, the $\Delta G°'$ is 0. Thus Equation 2-8 becomes

$$\Delta G = \Delta G°' + RT\, In Q = RT\, In \frac{C_2}{C_1}$$

where C_2 is the initial concentration of a substance inside the cell and C_1 is its concentration outside the cell. If the ratio of

C_2 to C_1 is 10, then at 25 °C, $\Delta G = RT \ln 10 = +1.36$ kcal per mole of substance transported. Such calculations assume that a molecule of a given substance inside a cell is identical with a molecule of that substance outside and that the substance is not sequestered, bound, or chemically changed by the transport.

Since the "uphill" transport of molecules against a concentration gradient ($C_2 > C_1$) has a positive ΔG, it clearly cannot take place spontaneously. To occur, such transport requires the input of cellular chemical energy, which often is supplied by the hydrolysis of ATP. Conversely, when a substance moves down its concentration gradient ($C_1 > C_2$) in crossing a membrane, ΔG has a negative value and the transport can be coupled to a reaction that has a positive ΔG, say, the movement of another substance uphill across a membrane.

Many Cellular Processes Involve Oxidation-Reduction Reactions

Many chemical reactions result in the transfer of electrons from one atom or molecule to another; this transfer may or may not accompany the formation of new chemical bonds. The loss of electrons from an atom or a molecule is called oxidation, and the gain of electrons by an atom or a molecule is called reduction. Because electrons are neither created nor destroyed in a chemical reaction, if one atom or molecule is oxidized, another must be reduced. For example, oxygen draws electrons from Fe^{2+} (ferrous) ions to form Fe^{3+} (ferric) ions, a reaction that occurs as part of the process by which carbohydrates are degraded in mitochondria. Each oxygen atom receives two electrons, one from each of two Fe^{2+} ions:

$$2Fe^{2+} + \tfrac{1}{2}O_2 \rightarrow 2\,Fe^{3+} + O^{2-}$$

Thus Fe^{2+} is oxidized, and O_2 is reduced. Oxygen similarly accepts electrons in many oxidation reactions in aerobic cells.

The transformation of succinate into fumarate is another oxidation reaction that takes place during carbohydrate breakdown in mitochondria. In this reaction, succinate loses two hydrogen atoms, which is equivalent to a loss of two protons and two electrons. Protons are soluble in aqueous solutions (as H_3O^+), but electrons are not and must be transferred directly from one atom or molecule to another. The electrons lost from succinate in its conversion to fumarate are transferred to flavin adenine dinucleotide (FAD), which is reduced to $FADH_2$. Many biologically important oxidation and reduction reactions involve the removal or the addition of hydrogen atoms (protons plus electrons) rather than the transfer of isolated electrons.

STANDARD REDUCTION POTENTIALS

To describe oxidation-reduction reactions, such as the reaction of ferrous ion (Fe^{2+}) and oxygen (O_2), it is easiest to divide them into two half-reactions:

Oxidation of Fe^{2+} : $2Fe^{2+} \rightarrow 2Fe^{3+} + 2e^-$

Reduction of O_2 : $2e^- + \frac{1}{2}O_2 \rightarrow O^{2-}$

In this case, the reduced oxygen (O^{2-}) readily reacts with two protons to form one water molecule:

$$2H^+ + O^{2-} \rightleftharpoons H_2O$$

Thus if we add two protons to each side of the equation for the half-reaction for reduction of O_2, the half-reaction can be rewritten as

$$2e^- + \frac{1}{2}O_2 + 2H^+ \rightleftharpoons H_2O$$

The readiness with which an atom or a molecule *gains* an electron is its reduction potential *E*. Reduction potentials are measured in volts (V) from an arbitrary zero point set at the reduction potential of the following half-reaction under standard conditions (25 °C, 1 atm, and reactants at 1 M):

$$H^{+} + e^{-} \underset{\text{oxidation}}{\overset{\text{reduction}}{\rightleftharpoons}} 1/2\ H_2$$

The value of *E* for a molecule or an atom under standard conditions is its standard reduction potential, E'_0. Standard reduction potentials may differ somewhat from those found under the conditions in a cell, because the concentrations of reactants in a cell are not 1 M. A positive reduction potential means that a molecule or ion (say, Fe^{3+}) has a higher affinity for electrons than the H^+ ion does in the standard reaction. A negative reduction potential means that a substance — for example, acetate (CH_3COO'') in its reduction to acetaldehyde (CH_3CHO) — has a lower affinity for electrons. In an oxidation-reduction reaction, electrons move spontaneously toward atoms or molecules having *more positive* reduction potentials. In other words, a compound having a more negative reduction potential (or more positive oxidation potential) can reduce — or transfer electrons to — one having a more positive reduction potential.

THE RELATIONSHIP BETWEEN CHANGES IN FREE ENERGY AND REDUCTION POTENTIALS

In an oxidation-reduction reaction, the total voltage change (change in electric potential) ΔE is the sum of the voltage changes (reduction potentials) of the individual oxidation or reduction steps. Because all forms of energy are interconvertible, we can express ΔE as a change in chemical free energy (ΔG). The charge in 1 mole (6.02×10^{23}) of electrons is 96,500 coulombs (96,500 joules per volt), a quantity known as the *Faraday constant* (f) after British physicist Michael Faraday (1791 – 1867). The following formula shows the relationship between free energy and reduction potential:

$$\Delta G(cal / mol) = -nf\ \Delta E = -n(96{,}500 / 4.184)\ \Delta E\,(volts)$$

or

$$\Delta G(cal / mol) = -n(23{,}064)\ \Delta E\,(volts) \qquad (2-11)$$

where *n* is the number of electrons transferred and 4.184 is the factor used to convert joules into calories. Note that an oxidation-reduction reaction with a positive ΔE value will have a negative ΔG and thus will tend to proceed from left to right.

The reduction potential is customarily used to describe the electric energy change that occurs when an atom or a molecule gains an electron. In an oxidation-reduction reaction, we also use the oxidation potential — the voltage change that takes place when an atom or molecule *loses* an electron — which is simply the negative of the reduction potential:

Reduction: $Cu^{2+} + e^- \rightarrow Cu^+ \qquad E'_0 = +0.35\,V$

Oxidation: $Cu^+ \rightarrow Cu^{2+} + e^- \qquad E'_0 = -0.35\,V$.

The voltage change in a complete oxidation-reduction reaction, in which one molecule is reduced and another is oxidized, is simply the sum of the oxidation potential and the reduction potential of the two partial oxidation and reduction reactions, respectively. Consider, for example, the change in electric potential (and, correspondingly, in standard free energy) when succinate is oxidized by oxygen:

$$Succinate + \tfrac{1}{2}\,O_2 \rightleftharpoons fumarate + H_2O$$

In this case, the partial reactions are

Oxidation:

Succinate $\rightleftharpoons$ $\qquad E'_0 = -0.03\ V$

Fumarate + $2H^+ + 2e^-$ (oxidation potential $\Delta G°' = +1.39$ kcal/mol (n=2 electrons)

Reduction:

$\tfrac{1}{2}O_2 + 2e^- + 2H^+ \rightleftharpoons$ $\qquad E'_0 = +\ 0.82\ V$

H_2O (reduction potential)

$\Delta G°' = -37.88$ kcal/mol

(n=2 electrons)

Sum: Succinate + $\frac{1}{2}\,O_2 \rightleftharpoons$ $E_0' =$ =

furmarate + H_2O −0.03 V + 0.82 V

= +0.79 V

$\Delta G°' =$

+1.39 kcal/mol−37.88kcal/mol

= −36.49 kcal/mol

The overall reaction has a positive $\Delta E'_0$ or, equivalently, a negative $\Delta G°'$ and thus, under standard conditions, will tend to occur from left to right.

An Unfavorable Chemical Reaction Can Proceed If It Is Coupled with an Energetically Favorable Reaction

Many chemical reactions in cells are energetically unfavorable ($\Delta G > 0$) and will not proceed spontaneously. One example is the synthesis of small peptides (e.g., glycylalanine) or proteins from amino acids. Cells are able to carry out a reaction that has a positive ΔG by coupling it to a reaction that has a negative ΔG of larger magnitude, so that the sum of the two reactions has a negative ΔG. Suppose that the reaction

$$A \rightleftharpoons B + X$$

has a $A \rightleftharpoons B + \Delta G°'$ of +5 kcal/mol and that the reaction

$$X \rightleftharpoons Y + Z$$

has a $\Delta G°'$ of "10 kcal/mol. In the absence of the second reaction, there would be much more A than B at equilibrium. The occurrence of the second process, by which *X* becomes *Y* + *Z*, changes that outcome: because it is such a favorable reaction, it will pull the first process toward the formation of B and the consumption of *A*.

The $\Delta G°'$ of the overall reaction will be the sum of the $\Delta G°'$ values of each of the two partial reactions:

$$A \rightleftharpoons B + X \qquad \Delta G°' = +5 \text{ kcl/mol}$$

$$X \rightleftharpoons Y + Z \qquad \Delta G°' = -10 \text{ kcl/mol}$$

Sum: $A \rightleftharpoons B + Y + Z$ $\Delta G^{\circ\prime} = -5$ kcl/mol

The overall reaction releases energy. In cells, energetically unfavorable reactions of the type $A \rightleftharpoons B + X$ are often coupled to the hydrolysis of the compound adenosine triphosphate (ATP), a reaction with a negative change in free energy ($\Delta G^{\circ\prime}$ = "7.3 kcal/mol), so that the overall reaction has a negative $\Delta G^{\circ\prime}$.

Hydrolysis of Phosphoanhydride Bonds in ATP Releases Substantial Free Energy

All cells extract energy from foods through a series of reactions that exhibit negative free-energy changes; plant cells also can extract energy from absorbed light. In both cases, much of the free energy is not allowed to dissipate as heat but is captured in chemical bonds formed by other molecules for use throughout the cell. In almost all organisms, the most important molecule for capturing and transferring free energy is adenosine triphosphate, or ATP.

The useful free energy in an ATP molecule is contained in phosphoanhydride bonds, which are formed from the condensation of two molecules of phosphate by the loss of water:

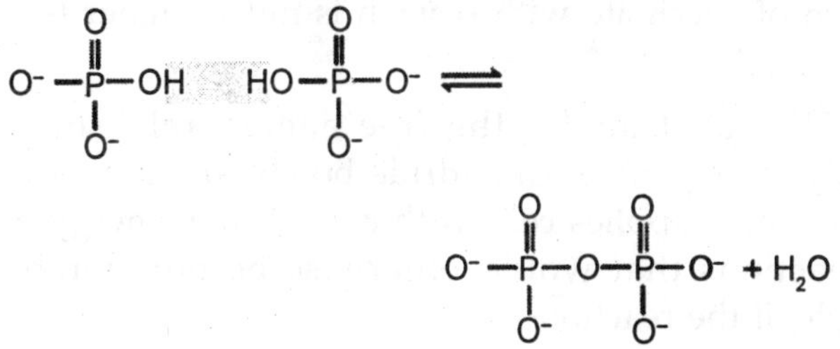

An *ATP* molecule has two phosphoanhydride bonds and is often written as adenosine – *p~p~p*, or simply *Ap~p~p*, where *p* stands for a phosphate group and ~ denotes a high-energy bond.

Hydrolysis of a phosphoanhydride bond in each of the following reactions has a highly negative $\Delta G^{\circ\prime}$ of about –7.3 kcal/mol:

$$Ap^{\sim}p^{\sim}p + H_2O \rightarrow Ap^{\sim}p + P_i + H^+$$
$$(ATP) \qquad (ADP)$$
$$Ap^{\sim}p^{\sim}p + H_2O \rightarrow Ap + PP_i + H^+$$
$$(ATP) \qquad (AMP)$$
$$Ap^{\sim}p + H_2O \rightarrow Ap + P_i + H^+$$
$$(ADP) \qquad (AMP)$$

In these reactions, P_i stands for inorganic phosphate and PP_i for inorganic pyrophosphate, two phosphate groups linked by a phosphoanhydride bond. As the top two reactions show, the removal of a phosphate or a pyrophosphate group from *ATP* leaves adenosine diphosphate (*ADP*) or adenosine monophosphate (*AMP*), respectively.

The phosphoanhydride bond is an ordinary covalent bond, but it releases about 7.3 kcal/mol of free energy (under standard biochemical conditions) when it is broken. In contrast, hydrolysis of the phosphoester bond in *AMP*, forming inorganic phosphate and adenosine, releases only about 2 kcal/mol of free energy. Phosphoanhydride bonds commonly are termed "high-energy" bonds, even though the $\Delta G^{\circ\prime}$ for the reaction of succinate with oxygen is much higher (–37 kcal/mol).

Cells can transfer the free energy released by the hydrolysis of phosphoanhydride bonds to other molecules. This transfer supplies cells with enough free energy to carry out reactions that would otherwise be unfavorable. For example, if the reaction

$$B + C \rightarrow D$$

is energetically unfavorable ($\Delta G > 0$), it can be made favorable by linking it to the hydrolysis of the terminal phosphoanhydride bond in *ATP*. Some of the energy in this phosphoanhydride bond is used to transfer a phosphate group to one of the reactants, forming a phosphorylated intermediate, *B~p*. The intermediate thus has enough free energy to react with *C*, forming *D* and free phosphate:

$$B + Ap^{\sim}p \rightarrow B^{\sim}p + Ap^{\sim}p$$

$$B^{\sim}p + C \rightarrow D + P_i$$

Thus, the overall reaction is

$$B + C + Ap^{\sim}p \sim p \rightarrow D^{\sim}Ap + p + P_i$$

which is energetically favorable. In the above example *B* and *C* would represent amino acids and *D* a dipeptide. Cells keep the ratio of ATP to ADP and AMP high, often as high as 10:1. Thus reactions in which the terminal phosphate group of ATP is transferred to another molecule will be driven even further along.

The $\Delta G^{o\prime}$ for hydrolysis of a phosphoanhydride bond in ATP (–7.3 kcal/mol) is about twice the $\Delta G^{o\prime}$ for hydrolysis of a phosphoester bond, such as that in glucose 6-phosphate (–3.3 kcal/mol). A principal reason for this difference is that ATP and its hydrolysis products ADP and P_i are highly charged at neutral *pH*. Three of the four ionizable protons in ATP are fully dissociated at pH 7.0, and the fourth, with a pK_a of 6.95, is about 50 percent dissociated. The closely spaced negative charges in ATP repel each other strongly. When the terminal phosphoanhydride bond is hydrolyzed, some of this stress is removed by the separation of the hydrolysis products ADP^{3-} and HPO_4^{2-}; that is, the separated negatively charged ADP^{3-} and HPO_4^{2-} will tend not to recombine to form ATP. In glucose 6-phosphate, by contrast, there is no charge repulsion between the phosphate group and the carbon atom to which it is attached. One of the hydrolysis products, glucose, is uncharged and will not repel the negatively charged HPO_4^{2-} ion; thus there is less resistance to the recombination of glucose and HPO_4^{2-} to form glucose 6-phosphate.

Many other bonds — particularly those between a phosphate group and some other substance — have the same high-energy character as phosphoanhydride bonds. The phosphoanhydride bond of ATP is not the most or the least energetic of these bonds. The preeminent role of ATP in capturing and transferring free energy within cells represents a compromise. The free energy of hydrolysis of ATP is sufficiently great that reactions in which the terminal phosphate group is transferred to another molecule have a substantially negative $\Delta G^{\circ\prime}$. However, if hydrolysis of this phosphoanhydride bond liberated considerably more free energy than it does, cells might require too much energy to form this bond in the first place. In other words, many reactions in cells release enough energy to form ATP, and hydrolysis of ATP releases enough energy to drive many of the cell's energy-requiring reactions and processes.

ATP IS USED TO FUEL MANY CELLULAR PROCESSES

If the terminal phosphoanhydride bond of ATP were to rupture by hydrolysis to produce ADP and P_i, energy would be released in the form of heat. However, cells contain various enzymes that can couple ATP hydrolysis to other reactions, so that much of the released energy is converted to more useful forms. For instance, cells use energy from ATP to synthesize macromolecules (proteins, nucleic acids, and polysaccharides) and many types of small molecules. The hydrolysis of ATP also supplies the energy needed to move individual cells from one location to another, to contract muscle cells, and to transport molecules into or out of the cell, usually against a concentration gradient. Gradients of ions, such as Na^+ and K^+, across a cellular membrane are produced by the action of membrane-embedded enzymes, called ion pumps, that couple the hydrolysis of ATP to the "uphill" movement of ions. The resulting ion concentration gradients are responsible for the generation of an electric potential across the membrane. This potential is the basis for the electric activity of cells and, in particular, for the conduction of impulses by nerves.

Clearly, to continue, functioning cells must constantly replenish their ATP supply. The ultimate energy source for formation of high-energy bonds in ATP and other compounds in nearly all cells is sunlight. Plants and microorganisms trap the energy in light through photosynthesis. In this process, chlorophyll pigments absorb the energy of light, which is then used to synthesize ATP from ADP and P_i. Much of the ATP produced in photosynthesis is used to help convert carbon dioxide to six-carbon sugars such as fructose and glucose:

$$6\ Co_2 + 6\ H_2O \xrightarrow{ATP \quad ATP + P_i} C_6H_{12}O_6 + 6O_2$$

Additional energy is used to convert hexoses into the disaccharide sucrose and polysaccharides. In animals, the free energy in sugars and other molecules derived from food is released in the process of respiration.

As noted earlier, glucose is a major source of energy in most cells. When 1 mole (180 g) of glucose reacts with oxygen under standard conditions according to the following reaction, 686 kcal of energy is released:

$$C_6H_{12}O_6 + 6O_2 \rightarrow 6CO_2 + 6H_2O$$

$$\Delta G^{o\prime} = -686 kcal/mol$$

If glucose is simply burned in air, all this energy is released as heat. By an elaborate set of enzyme-catalyzed reactions, cells couple the metabolism of 1 molecule of glucose to the synthesis of as many as 36 molecules of ATP from 36 molecules of ADP:

$$C_6H_{12}O_6 + 6O_2 + 36P_i + 36ADP \rightarrow 6CO_2 + 6H_2O + 36ATP$$

Because formation of one high-energy phosphoanhydride bond in *ATP*, from P_i and *ADP*, requires an input of 7.3 *kcal/mol*, about 263 *kcal* of energy (36 × 7.3) is conserved in *ATP* per mole of glucose metabolized (an efficiency of 263/686, or about 38 percent). This type of cellular metabolism is termed aerobic because it is dependent on the oxygen in the air. Aerobic catabolism (degradation) of glucose is found in all higher plant and animal cells and in many bacterial cells.

The overall reaction of glucose respiration

$$C_6H_{12}O_6 + 6O_2 \rightarrow 6CO_2 + 6H_2O$$

is the reverse of the photosynthetic reaction in which six-carbon sugars are formed

$$C_6H_{12}O_6 + 6H_2O \rightarrow C_6H_{12}O_6 + 6O_2$$

The latter reaction requires energy from light, whereas the former releases energy. Respiration and photosynthesis are the two major processes constituting the carbon cycle in nature: sugars and oxygen produced by plants are the raw materials for respiration and the generation of *ATP* by plant and animal cells alike; the end products of respiration, CO_2 and H_2O, are the raw materials for the photosynthetic production of sugars and oxygen. The only net source of energy in this cycle is sunlight. Thus, directly or indirectly, light energy captured in photosynthesis is the source of chemical energy for almost all cells.

The exceptions to this are certain microorganisms that exist in deep ocean vents where sunlight is completely absent. These unusual bacteria derive the energy for converting *ADP* and P_i into *ATP* from the oxidation of reduced inorganic compounds present in the dissolved vent gas that originates in the centre of the earth. Unfortunately, little is yet known about the biology of these organisms.

ACTIVATION ENERGY AND REACTION RATE

Many chemical reactions that exhibit a negative $\Delta G^{\circ\prime}$ do not proceed unaided at a measurable rate. For example, in pure aqueous solutions, glyceraldehyde 3-phosphate (*G3P*), our recurrent model, is a fairly stable compound that reacts very slowly or not at all, yet potentially it can undergo several different reactions, each of which has a negative $\Delta G^{\circ\prime}$. To understand why only one of several possible reactions generally occurs, we need to consider how reactions proceed and the factors that affect reaction rates.

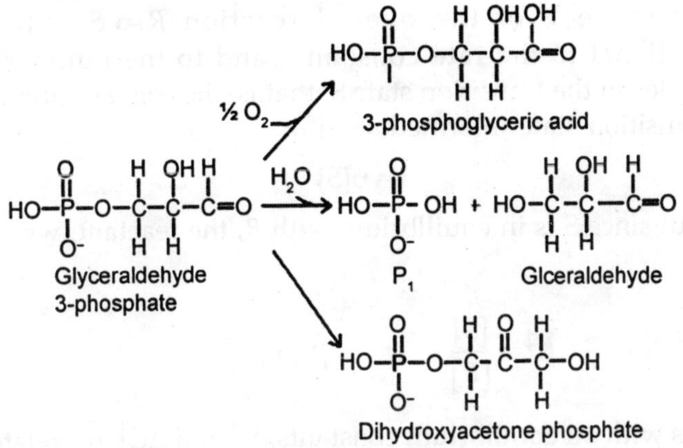

Fig. Glyceraldehyde 3-phosphate, like most cellular molecules, can undergo any of several reactions with negative ΔG°2 values: oxidation by O_2 to 3-phosphoglyceric acid, hydrolysis to glyceraldehyde and phosphate (P_i), and rearrangement to dihydroxyacetone phosphate.

CHEMICAL REACTIONS PROCEED THROUGH HIGH-ENERGY TRANSITION STATES

All chemical reactions proceed through one or more transition-state intermediates whose content of free energy is greater than that of either the reactants or the products. For the simple reaction R (reactants) $\rightleftharpoons$ P (products), we can write

$$R \overset{K^{\ddagger}}{\rightleftharpoons} S \xrightarrow{\upsilon} P$$

where S is the reaction intermediate with the highest free *energy; K‡ is the equilibrium constant for the reaction R $\rightleftharpoons$ S,* the conversion of the reactant to the high-energy intermediate S; and *v* is the rate constant for conversion of S into the product P. The free energy of activationΔG‡ is equal to the difference in free energy between the transition-state intermediate S and the reactant R. Because ΔG‡ generally has a very large positive value, only a small fraction of the reactant molecules will at any one time have acquired this free energy, and the overall rate of the reaction will be limited by the rate of formation of S.

The rate V of the overall reaction $R \rightarrow S$ will be proportional to the rate constant v and to the number of molecules in the transition state S, that is, the concentration of the transition-state intermediate, [S]:

$$V = \upsilon[S]$$

But since S is in equilibrium with R, the reactant, we can write

$$K^{\ddagger} = \frac{[S]}{[R]} \quad or \quad [S] = [R]K^{\ddagger}$$

As with all equilibrium constants, $K^{\ddagger}$ and $\Delta G^{\ddagger}$ are related as shown in Equation 2-10, so that

$$V = \upsilon[R] \times 10^{\left(\Delta G^{\ddagger}/2.3RT\right)}$$

From this equation, we can see that lowering the activation energy — that is, decreasing the free energy of the transition state $\Delta G^{\ddagger}$ — leads to an acceleration of the overall reaction rate V by increasing the concentration of S. A reduction in $\Delta G^{\ddagger}$ of 1.36 kcal/mol leads to a tenfold increase in the concentration of S, and thus a tenfold increase in the rate of the reaction. Similarly, a reduction of 2.72 *kcal/mol* in $\Delta G^{\ddagger}$ leads to a 100-fold increase in the reaction rate. Thus relatively small changes in $\Delta G^{\ddagger}$ can lead to large changes in the overall rate of the reaction.

In some reactions, certain covalent bonds are moved to a strained position in the transition state, and an input of energy — the energy of activation $\Delta G^{\ddagger}$ — is essential for this to happen. In other reactions, formation of the transition state involves excitation of electrons, which likewise requires an input of energy; only then can the electrons pair up, forming a covalent bond in the product. In still other reactions, molecules need only enough energy to overcome the mutual repulsion of their electron clouds to get close enough to react.

To illustrate the concept of a transition state we again consider the conversion of glyceraldehyde 3-phosphate (G3P) to dihydroxyacetone phosphate (DHAP), which involves at least one reaction intermediate. As the intermediate forms, the following events take place simultaneously: a proton is removed from carbon 2 of G3P, another proton is donated to the aldehyde oxygen on carbon 1, and pairs of electrons move from one bond to another. The activation energy required by each of these events contributes to the overall activation energy needed to form this reaction intermediate, which then rearranges through a second transition-state intermediate to generate the final reaction product. Each stage in such a multistep reaction has its own activation energy but for the overall reaction to proceed, the highest activation energy must be achieved.

At room or body temperature, the average kinetic energy, the energy of motion, of a typical molecule is about 1.5 kcal/mol. Although many molecules will have more kinetic energy than this average, the kinetic energy of colliding molecules is generally insufficient to provide the necessary activation energy to convert a reactant to the transition state and thus to allow a particular reaction to proceed. Without some mechanism for accelerating reactions, cells would be able to carry out few, if any, of the biochemical reactions needed to sustain life.

ENZYMES ACCELERATE BIOCHEMICAL REACTIONS BY REDUCING TRANSITION-STATE FREE ENERGY

Two significant rate-regulating factors for biological systems are the concentrations of the reactants and the pH. A reaction involving two or more different molecules proceeds faster at high concentrations because the molecules are more likely to encounter one another. The pH determines the dissociation state of the various acidic and basic groups on biological molecules. Since only one of the possible ionic forms (e.g., the zwitterion form of an amino acid) can undergo a particular reaction, the pH can determine the amount of the reactive species present. However, the most important determinants of biochemical reaction rates are enzymes, the proteins that act as catalysts. The structure and mechanism of action of enzymes are discussed in detail in the next chapter. Here we examine their general effect on reactions.

As discussed earlier, enzymes, like all catalysts, cause reactions to reach equilibrium faster. A catalyst accelerates the rates of forward and reverse reactions by the same factor; it does not alter the change in free energy or the equilibrium constant. Enzymes and all other catalysts act by reducing the activation energy required to make a reaction proceed. To achieve this, an enzyme binds either a single substrate or a set of similar substrates. Each enzyme catalyzes a single chemical reaction on the bound substrate. Triosephosphate isomerase, the enzyme that catalyzes the conversion of G3P to DHAP, binds G3P as its substrate. Once G3P is bound, the requisite movements of its hydrogen atoms, protons, and electrons are all facilitated by specific chemical groups on the parts of the enzyme adjacent to the bound substrate.

Some enzymes bind a substrate in a way that strains certain of its bonds and makes it easy for these bonds in the substrate to undergo a reaction. That is, the enzyme stabilizes the transition state of the reaction by tightly binding a specific form of the substrate in which certain bonds are strained. Other enzymes form a covalent bond with the substrate that enables a different part of the substrate to undergo a reaction; after this happens, the bond between the enzyme and substrate is broken. Still other enzymes bind multiple substrates in a way that brings them close enough so they can react readily with one another. In each case, the overall effect is to reduce the activation energy needed for formation of the reaction intermediate. Thus the presence or absence of particular enzymes in a cell or in extracellular fluids determines which of many possible chemical reactions will occur.

Chapter 3

The Proteins

Proteins, the working molecules of a cell, carry out the programme of activities encoded by genes. This programme requires the coordinated effort of many different types of proteins, which first evolved as rudimentary molecules that facilitated a limited number of chemical reactions. Gradually, many of these primitive proteins evolved into a wide array of enzymes capable of catalyzing an incredible range of intracellular and extracellular chemical reactions, with a speed and specificity that is nearly impossible to attain in a test tube. Other proteins acquired numerous structural, regulatory, and other functions. For a flavor of the various roles of proteins in today's organisms, we can look to the yeast *Saccharomyces cerevisiae,* a simple unicellular eukaryote. The yeast genome is predicted to encode about 6225 proteins. On the basis of their sequences, 17 percent are estimated to be involved in metabolism, the synthesis or degradation of cell building blocks; 30 percent, in cellular organization and biogenesis of cell organelles and membranes; and 10 percent, in transporting molecules across membranes.

In this chapter, we will study how the structure of a protein gives rise to its function. The first section examines protein architecture: the structure and chemistry of amino acids, the linkage of amino acids to form a linear chain, and the forces that guide folding of the chain into higher orders of structure. In the next section, we learn about special proteins that aid in the folding of proteins, modifications that occur after the protein chain is synthesized, and mechanisms that degrade proteins. In the third section, we illustrate several key concepts in the functional design of proteins, using antibodies and enzymes as examples. A separate section is devoted to the general characteristics of membrane proteins, which reside in the lipid bilayer surrounding cells and organelles. These functionally diverse proteins play critical roles in transfer of molecules and information across the lipid bilayer and in cell-cell interactions; their structures and functions will be discussed in greater detail in later chapters. We finish the chapter by describing the most commonly used techniques in the biologist's tool kit for isolating proteins and characterizing their properties. Our understanding of biology critically depends on how we can ask a question and test it experimentally.

HIERARCHICAL STRUCTURE OF PROTEINS

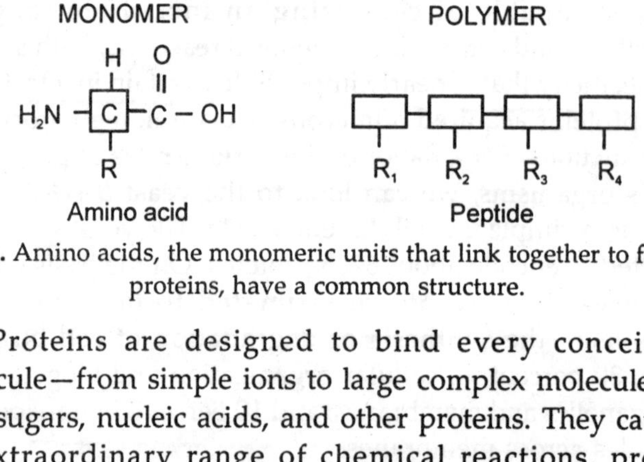

Fig. Amino acids, the monomeric units that link together to form proteins, have a common structure.

Proteins are designed to bind every conceivable molecule—from simple ions to large complex molecules like fats, sugars, nucleic acids, and other proteins. They catalyze an extraordinary range of chemical reactions, provide structural rigidity to the cell, control flow of material through membranes, regulate the concentrations of metabolites, act as sensors and switches, cause motion, and control gene function. The three-dimensional structures of proteins have evolved to carry out these functions efficiently and under precise control. The *spatial* organization of proteins, their shape in three dimensions, is a key to understanding how they work.

One of the major areas of biological research today is how proteins, constructed from only 20 different amino acids, carry out the incredible array of diverse tasks that they do. Unlike the intricate branched structure of carbohydrates, proteins are single, unbranched chains of amino acid monomers. The unique shape of proteins arises from noncovalent interactions between regions in the linear sequence of amino acids. Only when a protein is in its correct three-dimensional structure, or conformation, is it able to function efficiently. A key concept in understanding how proteins work is that *function is derived from three-dimensional structure, and three-dimensional structure is specified by amino acid sequence.*

THE AMINO ACIDS COMPOSING PROTEINS DIFFER ONLY IN THEIR SIDE CHAINS

Amino acids are the monomeric building blocks of proteins. The α carbon atom (C_α) of amino acids, which is adjacent to the carboxyl group, is bonded to four different chemical groups: an amino (NH_2) group, a carboxyl (COOH) group, a hydrogen (H) atom, and one variable group, called a *side chain* or *R group*. All 20 different amino acids have this same general structure, but their side-chain groups vary in size, shape, charge, hydrophobicity, and reactivity.

The amino acids can be considered the alphabet in which linear proteins are "written." Students of biology must be familiar with the special properties of each letter of this alphabet, which are determined by the side chain. Amino acids can be classified into a few distinct categories based primarily on their solubility in water, which is influenced by the polarity of their side chains. Amino acids with polar side groups tend to be on the surface of proteins; by interacting with water, they make proteins soluble in aqueous solutions. In contrast, amino acids with nonpolar side groups avoid water and aggregate to form the waterinsoluble core of proteins. The polarity of amino acid side chains thus is one of the forces responsible for shaping the final three-dimensional structure of proteins.

Hydrophilic, or water-soluble, amino acids have ionized or polar side chains. At neutral pH, *arginine* and *lysine* are positively charged; *aspartic acid* and *glutamic acid* are negatively charged and exist as aspartate and glutamate. These four amino acids are the prime contributors to the overall charge of a protein. A fifth amino acid, *histidine*, has an imidazole side chain, which has a pK_a of 6.8, the pH of the cytoplasm. As a result, small shifts of cellular pH will change the charge of histidine side chains:

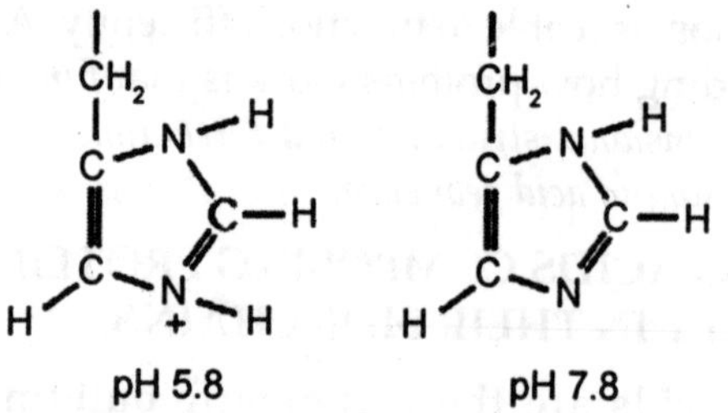

The activities of many proteins are modulated by pH through protonation of histidine side chains. *Asparagine* and *glutamine* are uncharged but have polar amide groups with extensive hydrogen-bonding capacities. Similarly, *serine* and *threonine* are uncharged but have polar hydroxyl groups, which also participate in hydrogen bonds with other polar molecules. Because the charged and polar amino acids are hydrophilic, they are usually found at the surface of a water-soluble protein, where they not only contribute to the solubility of the protein in water but also form binding sites for charged molecules.

Hydrophobic amino acids have aliphatic side chains, which are insoluble or only slightly soluble in water. The side chains of *alanine, valine, leucine, isoleucine,* and *methionine* consist entirely of hydrocarbons, except for the sulfur atom in methionine, and all are nonpolar. *Phenylalanine, tyrosine,* and *tryptophan* have large bulky aromatic side groups. As explained hydrophobic molecules avoid water by coalescing into an oily or waxy droplet. The same forces cause hydrophobic amino acids to pack in the interior of proteins, away from the aqueous environment. Later in this chapter, we will see in detail how hydrophobic residues line the *surface* of membrane proteins that reside in the hydrophobic environment of the lipid bilayer.

Lastly, *cysteine, glycine,* and *proline* exhibit special roles in proteins because of the unique properties of their side chains. The side chain of cysteine contains a reactive sulfhydryl group (—SH), which can oxidize to form a disulfide bond (—S—S—) to a second cysteine:

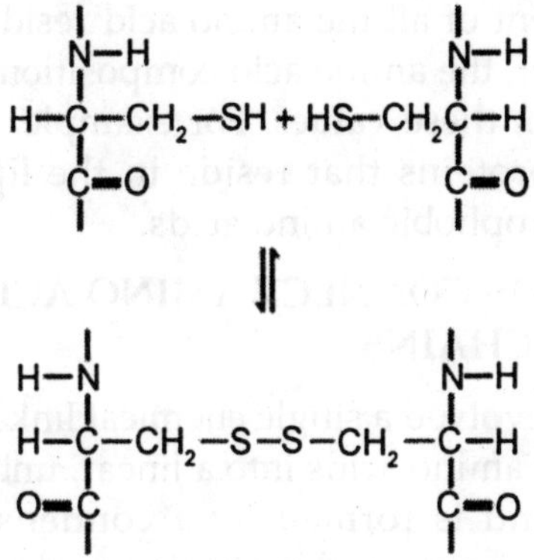

Regions within a protein chain or in separate chains sometimes are cross-linked covalently through disulfide bonds. Although disulfide bonds are rare in intracellular proteins, they are commonly found in extracellular proteins, where they help maintain the native, folded structure. The smallest amino acid, glycine, has a single hydrogen atom as its R group. Its small size allows it to fit into tight spaces. Unlike any of the other common amino acids, proline has a cyclic ring that is produced by formation of a covalent bond between its R group and the amino group on C_α. Proline is very rigid, and its presence creates a fixed kink in a protein chain. Proline and glycine are sometimes found at points on a protein's surface where the chain loops back into the protein.

The 6225 known and predicted proteins encoded by the yeast genome have an average molecular weight (MW) of 52,728 and contain, on average, 466 amino acid residues. Assuming that these average values represent a "typical" eukaryotic protein, then the average molecular weight of amino acids is 113, taking their average relative abundance in proteins into account. This is a useful number to remember, as we can use it to estimate the number of residues from the molecular weight of a protein or vice versa. Some amino acids are more abundant in proteins than other amino acids. Cysteine, tryptophan, and methionine are rare amino acids; together they constitute approximately 5 percent of the amino acids in a protein. Four amino acids—leucine, serine, lysine, and glutamic acid—are the most abundant amino acids, totaling 32 percent of all the amino acid residues in a typical protein. However, the amino acid composition of proteins can vary widely from these values. For example, as discussed in later sections, proteins that reside in the lipid bilayer are enriched in hydrophobic amino acids.

PEPTIDE BONDS CONNECT AMINO ACIDS INTO LINEAR CHAINS

Nature has evolved a single chemical linkage, the peptide bond, to connect amino acids into a linear, unbranched chain. The peptide bond is formed by a condensation reaction between the amino group of one amino acid and the carboxyl group of another. The repeated amide N, C_α, and carbonyl C atoms of each amino acid residue form the backbone of a protein molecule from which the various side-chain groups project. As a consequence of the peptide linkage, the backbone has polarity, since all the amino groups lie to the same side of the C_α atoms. This leaves at opposite ends of the chain a free (unlinked) amino group (the N-terminus) and a free carboxyl group (the C-terminus). A protein chain is conventionally depicted with its N-terminal amino acid on the left and its C-terminal amino acid on the right.

Many terms are used to denote the chains formed by polymerization of amino acids. A short chain of amino acids linked by peptide bonds and having a defined sequence is a peptide; longer peptides are referred to as polypeptides. Peptides generally contain fewer than 20–30 amino acid residues, whereas polypeptides contain as many as 4000 residues. We reserve the term protein for a polypeptide (or a complex of polypeptides) that has a threedimensional structure. It is implied that proteins and peptides represent natural products of a cell.

The size of a protein or a polypeptide is reported as its mass in daltons (a dalton is 1 atomic mass unit) or as its molecular weight (a dimensionless number). For example, a 10,000-MW protein has a mass of 10,000 daltons (Da), or 10 kilodaltons (kDa). In the last section of this chapter, we will discuss different methods for measuring the sizes and other physical characteristics of proteins.

FOUR LEVELS OF STRUCTURE DETERMINE THE SHAPE OF PROTEINS

The structure of proteins commonly is described in terms of four hierarchical levels of organization. These levels which depicts the structure of hemagglutinin, a surface protein on the influenza virus. This protein binds to the surface of animal cells, including human cells, and is responsible for the infectivity of the flu virus.

The primary structure of a protein is the linear arrangement, or *sequence,* of amino acid residues that constitute the polypeptide chain.

Secondary structure refers to the localized organization of parts of a polypeptide chain, which can assume several different spatial arrangements. A single polypeptide may exhibit all types of secondary structure. Without any stabilizing interactions, a polypeptide assumes a *random-coil* structure. However, when stabilizing hydrogen bonds form between certain residues, the backbone folds periodically into one of two geometric arrangements: an α helix, which is a spiral, rodlike structure, or a β sheet, a planar structure composed of alignments of two or more β *strands,* which are relatively short, fully extended segments of the backbone. Finally, U-shaped four-residue segments stabilized by hydrogen bonds between their arms are called *turns.* They are located at the surfaces of proteins and redirect the polypeptide chain toward the interior. (These structures will be discussed in greater detail later.)

Tertiary structure, the next-higher level of structure, refers to the overall conformation of a polypeptide chain, that is, the three-dimensional arrangement of all the amino acids residues. In contrast to secondary structure, which is stabilized by hydrogen bonds, tertiary structure is stabilized by hydrophobic interactions between the nonpolar side chains and, in some proteins, by disulfide bonds. These stabilizing forces hold the α helices, β strands, turns, and random coils in a compact internal scaffold. Thus, a protein's size and shape is dependent not only on its sequence but also on the number, size, and arrangement of its secondary structures. For proteins that consist of a single polypeptide chain, monomeric proteins, tertiary structure is the highest level of organization.

Multimeric proteins contain two or more polypeptide chains, or *subunits*, held together by noncovalent bonds. Quaternary structure describes the number (stoichiometry) and relative positions of the subunits in a multimeric protein. Hemagglutinin is a trimer of three identical subunits; other multimeric proteins can be composed of any number of identical or different subunits.

In a fashion similar to the hierarchy of structures that make up a protein, proteins themselves are part of a hierarchy of cellular structures. Proteins can associate into larger structures termed *macromolecular assemblies*. Examples of such macromolecular assemblies include the protein coat of a virus, a bundle of actin filaments, the nuclear pore complex, and other large submicroscopic objects. Macromolecular assemblies in turn combine with other cell biopolymers like lipids, carbohydrates, and nucleic acids to form complex cell organelles.

GRAPHIC REPRESENTATIONS OF PROTEINS HIGHLIGHT DIFFERENT FEATURES

Different ways of depicting proteins convey different types of information. The simplest way to represent three-dimensional structure is to trace the course of the backbone atoms with a solid line ; the most complex model shows the location of every atom. The former shows the overall organization of the polypeptide chain without consideration of the amino acid side chains; the latter details the interactions among atoms that form the backbone and that stabilize the protein's conformation. Even though both views are useful, the elements of secondary structure are not easily discerned in them.

Another type of representation uses common shorthand symbols for depicting secondary structure, cylinders for α helices, arrows for β strands, and a flexible stringlike form for parts of the backbone without any regular structure. This type of representation emphasizes the organization of the secondary structure of a protein, and various combinations of secondary structures are easily seen.

However, none of these three ways of representing protein structure conveys much information about the protein surface, which is of interest because this is where other molecules bind to a protein. Computer analysis in which a water molecule is rolled around the surface of a protein can identify the atoms that are in contact with the watery environment. On this water-accessible surface, regions having a common chemical (hydrophobicity or hydrophilicity) and electrical (basic or acidic) character can be mapped. Such models show the texture of the protein surface and the distribution of charge, both of which are important parameters of binding sites. This view represents a protein as seen by another molecule.

SECONDARY STRUCTURES ARE CRUCIAL ELEMENTS OF PROTEIN ARCHITECTURE

In an average protein, 60 percent of the polypeptide chain exists as two regular secondary structures, α helices and β sheets; the remainder of the molecule is in random coils and turns. Thus, α helices and β sheets are the major internal supportive elements in proteins. In this section, we explore the forces that favour formation of secondary structures. In later sections, we examine how these structures can pack into larger arrays.

THE α HELIX

Polypeptide segments can assume a regular spiral, or

helical, conformation, called the α *helix.* In this secondary structure, the carbonyl oxygen of each peptide bond is hydrogen-bonded to the amide hydrogen of the amino acid four residues toward the C-terminus. This uniform arrangement of bonds confers a polarity on a helix because all the hydrogen-bond donors have the same orientation. The peptide backbone twists into a helix having 3.6 amino acids per turn. The stable arrangement of amino acids in the α helix holds the backbone as a rodlike cylinder from which the side chains point outward. The hydrophobic or hydrophilic quality of the helix is determined entirely by the side chains, because the polar groups of the peptide backbone are already involved in hydrogen bonding in the helix and thus are unable to affect its hydrophobicity or hydrophilicity.

In many α helices hydrophilic side chains extend from one side of the helix and hydrophobic side chains from the opposite side, making the overall structure amphipathic. In such helices the hydrophobic residues, although apparently randomly arranged, occur in a regular pattern. One way of visualizing this arrangement is to look down the centre of an α helix and then project the amino acid residues onto the plane of the paper. The residues will appear as a wheel, and in the case of an amphipathic helix, the hydrophobic residues all lie on one side of the wheel and the hydrophilic ones on the other side.

Amphipathic α helices are important structural elements in fibrous proteins found in a watery environment. In a coiled-coil region of a protein, the hydrophobic surface of the α helix faces inward to form the hydrophobic core, and the hydrophilic surfaces face outward toward the surrounding fluid. This same orientation of surfaces is also found in most globular proteins. A crucial difference is that the hydrophobic interaction could be with a β strand, random coil, or another α helix. As we discuss later, amphipathic β strands line the walls of an ion channel in the cell membrane.

THE β SHEET

Another regular secondary structure, the β sheet, consists

of laterally packed β strands. Each β strand is a short (5–8-residue), nearly fully extended polypeptide chain. Hydrogen bonding between backbone atoms in adjacent β strands, within either the same or different polypeptide chains, forms a β sheet. Like α helices, β strands have a polarity defined by the orientation of the peptide bond. Therefore, in a pleated sheet, adjacent β strands can be oriented antiparallel or parallel with respect to each other. In both arrangements of the backbone, the side chains project from both faces of the sheet.

In some proteins, β sheets form the floor of a binding pocket. In many structural proteins, multiple layers of pleated sheets provide toughness. Silk fibers, for example, consist almost entirely of stacks of antiparallel β sheets. The fibers are flexible because the stacks of β sheets can slip over one another. However, they are also resistant to breakage because the peptide backbone is aligned parallel with the fiber axis.

TURNS

Composed of three or four residues, turns are compact, U-shaped secondary structures stabilized by a hydrogen bond between their end residues. They are located on the surface of a protein, forming a sharp bend that redirects the polypeptide backbone back toward the interior. Glycine and proline are commonly present in turns. The lack of a large side chain in the case of glycine and the presence of a built-in bend in the case of proline allow the polypeptide backbone to fold into a tight U-shaped structure. Without turns, a protein would be large, extended, and loosely packed. A polypeptide backbone also may contain long bends, or *loops*. In contrast to turns, which exhibit a few defined structures, loops can be formed in many different ways.

MOTIFS ARE REGULAR COMBINATIONS OF SECONDARY STRUCTURES

Many proteins contain one or more motifs built from particular combinations of secondary structures. A motif is defined by a specific combination of secondary structures that has a particular topology and is organized into a characteristic three-dimensional structure.

The coiled-coil motif comprises two, three, or four amphipathic α helices wrapped around one another. In this motif, hydrophobic side chains project like "knobs" from one helix and interdigitate into the gaps, or "holes," between the hydrophobic side chains of the other helix along the contact surface. The subunits in some multimeric proteins and in rodlike fibers are held together by coiled-coil interactions. The Ca^{2+}-binding helix-loop-helix motif is marked by the presence of certain hydrophilic residues at invariant positions in the loop. Oxygen atoms in the invariant residues bind a calcium ion through hydrogen bonds. In another common motif, the zinc finger, three secondary structures—an α helix and two β strands with an antiparallel orientation—form a fingerlike bundle held together by a zinc ion. This motif is most commonly found in proteins that bind RNA or DNA.

Additional motifs will be examined in discussions of other proteins. The presence of the same motif in different proteins with similar functions clearly indicates that during evolution these useful combinations of secondary structures have been conserved.

STRUCTURAL AND FUNCTIONAL DOMAINS ARE MODULES OF TERTIARY STRUCTURE

The tertiary structure of large proteins is often subdivided into distinct globular or fibrous regions called domains. Structurally, a domain is a compactly folded region of polypeptide. For large proteins, domains can be recognized in structures determined by x-ray crystallography or in images captured by electron microscopy. These discrete regions are well distinguished or physically separated from other parts of the protein, but connected by the polypeptide chain. Hemagglutinin, for example, contains a globular domain and a fibrous domain.

A structural domain consists of 100–200 residues in various combinations of α helices, β sheets, turns, and random coils. Often a domain is characterized by some interesting structural feature, for example, an unusual abundance of a particular amino acid (a proline-rich domain, an acidic domain, a glycine-rich domain), sequences common to (conserved in) many proteins (SH3, or Src homology region 3), or a particular secondary-structure motif (zinc-finger motif in kringle domain).

Domains sometimes are defined in functional terms based on observations that the activity of a protein is localized to a small region along its length. For instance, a particular region or regions of a protein may be responsible for its catalytic activity (e.g., a kinase domain) or binding ability (e.g., a DNA-binding domain, membrane-binding domain). Functional domains often are identified experimentally by whittling down a protein to its smallest active fragment with the aid of proteases, enzymes that cleave the polypeptide backbone. Alternatively, the DNA encoding a protein can be subjected to mutagenesis, so that segments of the protein's backbone are removed or changed. The activity of the truncated or altered protein product synthesized from the mutated gene is then monitored.

The functional definition of a domain is less rigorous than a structural definition. However, if the three-dimensional structure of a protein has not been determined, identification of functional domains can provide useful information about the protein. Because the activity of a protein usually depends on a proper three-dimensional structure, a functional domain consists of at least one and often several structural domains.

The organization of tertiary structure into domains further illustrates the principle that complex molecules are built from simpler components. Like secondary-structure motifs, tertiary-structure domains are incorporated as modules into different proteins, thereby modifying their functional activities. The modular approach to protein architecture is particularly easy to recognize in large proteins, which tend to be a mosaic of different domains and thus can perform different functions simultaneously.

The epidermal growth factor (EGF) domain is one example of a module that is present in several proteins. EGF is a small soluble peptide hormone that binds to cells in the skin and connective tissue, causing them to divide. It is generated by proteolytic cleavage between repeated EGF domains in the EGF precursor protein, which is anchored in the cell membrane by a membrane-spanning domain. Six conserved cysteine residues form three pairs of disulfide bonds that hold EGF in its native conformation. The EGF domain also occurs in other proteins, including tissue plasminogen activator (TPA), a protease that is used to dissolve blood clots in heart attack victims; Neu protein, which is involved in embryonic differentiation; and Notch protein, a cell-adhesion molecule that glues cells together. Besides the EGF domain, these proteins contain additional domains found in other proteins. For example, TPA possesses a chymotryptic domain, a common feature in proteins that catalyze proteolysis.

SEQUENCE HOMOLOGY SUGGESTS FUNCTIONAL AND EVOLUTIONARY RELATIONSHIPS BETWEEN PROTEINS

Early evidence supporting the key principle that the amino acid sequence of a protein determines its three-dimensional structure was obtained in the 1960s by Max Perutz. On comparing the structures of myoglobin and hemoglobin determined from x-ray crystallographic analysis, he immediately noted that the subunits of hemoglobin, a tetramer of two α and two β subunits, resembled myoglobin, a monomer. Although the sequences of the two proteins were unknown at the time, Perutz proposed that the similar arrangement of α helices in the two proteins is a consequence of their having similar amino acid sequences. Later sequencing of myoglobin and hemoglobin revealed that many identical or chemically similar residues occur in identical positions throughout the sequences of both proteins. The two proteins also exhibit similar functions: myoglobin is the oxygen-carrier protein in muscle, and hemoglobin the oxygen-carrier protein in blood. Most of the conserved residues hold the heme group in place or are responsible for maintaining the hydrophobic interior of the protein.

As data concerning protein sequences and three-dimensional structures accumulated, the concept that similar sequences fold into similar secondary and tertiary structures was confirmed. The propensity of each amino acid to occur in the various types of secondary structures has been calculated from the amino acid sequence of secondary structures extracted from databases of the three-dimensional structures of proteins. This tabulation of the folding information inherent in the sequence is now being used in attempts to predict the three-dimensional structure of various proteins from their amino acid sequences.

In the classical taxonomy of the eighteenth and nineteenth centuries, organisms were classified according to their morphological similarities and differences. In this century, the molecular revolution in biology has given birth to "molecular" taxonomy: the classification of proteins based on similarities and differences in their amino acid sequences. This new taxonomy provides much information about protein function and evolutionary relationships. If the similarity between proteins from different organisms is significant over their entire sequence, then the proteins are homologs of one another, and they probably carry out similar functions. Sequence similarity also suggests an evolutionary relationship between proteins; that is, they evolved from a common ancestor. We can therefore describe homologous proteins as belonging to the same "family" and can trace their lineage from comparisons of sequences. Closely related proteins have the most similar sequences; distantly related proteins have only faintly similar sequences.

The kinship among homologous proteins is most easily visualized from a tree diagram based on sequence analyses. For example, the amino acid sequences of hemoglobins from different species suggest that they evolved from an ancestral monomeric, oxygen-binding protein. Over time, this ancestral protein slowly changed, giving rise to myoglobin, which remained a monomeric protein, and to the α and β subunits, which evolved to associate into the tetrameric hemoglobin molecule. Evolution of the globin protein family parallels that of the vertebrates.

The power of such comparative analysis and identification of homologous proteins has expanded substantially in recent years by use of the base sequences in an organism's genome to deduce the amino acid sequences of the encoded proteins.

FOLDING, MODIFICATION, AND DEGRADATION OF PROTEINS

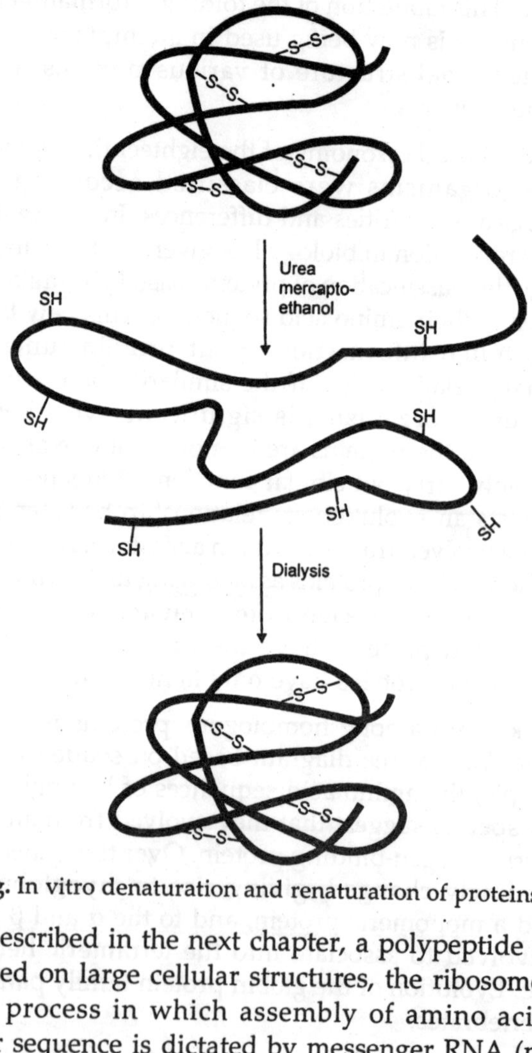

Fig. In vitro denaturation and renaturation of proteins.

As described in the next chapter, a polypeptide chain is synthesized on large cellular structures, the ribosomes, by a complex process in which assembly of amino acids in a particular sequence is dictated by messenger RNA (mRNA). The nascent polypeptide chain undergoes folding and, in many cases, chemical modification to generate the final protein. Any polypeptide chain containing *n* residues could, in principle, fold into 8^n conformations. This value is based on the fact that only eight bond angles are stereochemically allowed in the polypeptide backbone. In general, however, all molecules of any protein species adopt a single conformation, called the *native state,* which is the most stably folded form of the molecule. Misfolding to non-native conformations is suppressed by two distinct mechanisms. At the molecular level, a protein folds through a pathway that favors only a few intermediate steps. Furthermore, a cellular system prevents misfolded proteins from forming. After a protein has carried out its functions, specific sequences that limit the life span of the protein target it for degradation.

THE INFORMATION FOR PROTEIN FOLDING IS ENCODED IN THE SEQUENCE

The realization that the amino acid sequence of a protein determines its folding came from in vitro studies on protein unfolding and refolding. Thermal energy from heat, extremes of pH that alter the charges on amino acid side chains, and chemicals such as urea or guanidine hydrochloride at concentrations of 6–8 M can disrupt the weak noncovalent bonds that stabilize the native conformation of a protein. The denaturation resulting from such treatment causes a protein to lose both its compact conformation and activity. Most denatured proteins precipitate in solution because hydrophobic groups, normally buried inside the molecules, interact with similar regions of other unfolded molecules, causing them to form an insoluble aggregate.

Many proteins that are completely unfolded in 8 M urea and β-mercaptoethanol (which reduces disulfide bonds) can *renature* (refold) into their native state when the denaturing reagents are removed by dialysis. During renaturation, all the disulfide, hydrogen, and hydrophobic bonds that stabilize the native conformation are re-formed. Thus, in this case proteins can be carried through a denaturationrenaturation cycle, which first destroys and then reestablishes their original structure and function. Because renaturation requires no cofactors or other proteins, at least in the test tube, protein folding is a self-assembly process.

The observation by Christian Anfinsen of such reversible denaturation and renaturation of ribonuclease, an enzyme that degrades RNA, provided a clue that the information for folding a protein lies in its sequence. A general mechanism by which proteins refold in vitro has been elucidated in experiments in which the renaturing conditions are carefully adjusted and the refolding reaction is interrupted at various time intervals. Such studies have shown that the polypeptide goes through several transient reconfigurations, including a "molten globule" state, before the native tertiary conformation is reached. In the case of ribonuclease, which has several internal disulfide bonds, the folding pathway involves rearrangements of disulfide bond pairs to the native conformation.

FOLDING OF PROTEINS IN VIVO IS PROMOTED BY CHAPERONES

Folding of proteins in vitro is an inefficient process, with only a minority of unfolded molecules undergoing complete folding within a few minutes. Clearly, in vivo most protein molecules must rapidly fold into their correct shape; otherwise, cells would waste much energy in the synthesis of nonfunctional proteins and in the degradation of misfolded or unfolded proteins. More than 95 percent of the proteins present within cells have been shown to be in their native conformation, despite high protein concentrations (≈100 mg/ml), which usually cause proteins to precipitate in vitro.

The explanation for the cell's remarkable efficiency in promoting protein folding probably lies in chaperones, a family of proteins found in all organisms from bacteria to humans. Chaperones are located in every cellular compartment, bind a wide range of proteins, and may be part of a general protein-folding mechanism. There are two general families of chaperones: *molecular chaperones*, which bind and stabilize unfolded or partially folded proteins, thereby preventing these proteins from being degraded; and *chaperonins,* which directly facilitate their folding. Chaperones have ATPase activity, and their ability to bind and stabilize their target proteins is specific and dependent on ATP hydrolysis. Binding of chaperones to partially folded proteins suggests that the folding process could be regulated at intermediate steps.

Molecular chaperones consist of the Hsp70 family of proteins, which includes Hsp70 in the cytosol and mitochondrial matrix, Bip in the endoplasmic reticulum, and DnaK, a bacterial chaperone. First identified by its rapid appearance after a cell has been stressed by heat shock, Hsp70 is the major chaperone protein in all organisms. When bound to ATP, Hsp70 assumes an open form in which an exposed hydrophobic pocket transiently binds to exposed hydrophobic regions of the unfolded target protein. Hydrolysis of the bound ATP causes Hsp70 to assume a closed form, releasing the target protein. Molecular chaperones are thought to bind all nascent polypeptide chains as they are being synthesized on ribosomes. In bacteria, 85 percent of the proteins are released from their chaperone and go on to fold normally; an even higher percentage of proteins in eukaryotes follow this pathway.

Proper folding of a small proportion of proteins (e.g., the cytoskeletal proteins actin and tubulin) requires additional assistance, which is provided by chaperonins. Eukaryotic chaperonins, called *TCiP,* are large, barrel-shaped, multimeric complexes composed of eight Hsp60 units. The bacterial homolog, known as *GroEL,* contains 14 identical subunits. The GroEL folding mechanism, which is better understood than TCiP-mediated folding, serves as a reasonable general model. In bacteria, a partially folded or misfolded polypeptide is inserted into the cavity of GroEL, where it binds to the inner wall and folds into its native conformation. In an ATP-dependent step, GroEL expands and the protein exits GroEL, a process assisted by a co-chaperonin, GroES, which caps the ends of GroEL. Because the eukaryotic chaperonin TCiP lacks a GroES-type co-chaperonin, the last step must differ in eukaryotes. Moreover, the size of the cavity in TCiP limits this folding pathway to polypeptides smaller than 55 kDa.

CHEMICAL MODIFICATIONS AND PROCESSING ALTER THE BIOLOGICAL ACTIVITY OF PROTEINS

Nearly every protein in a cell is chemically altered after its synthesis on a ribosome. Such modifications may alter the activity, life span, or cellular location of proteins, depending on the nature of the alteration. Protein alterations fall into two categories: chemical modification and processing. *Chemical modification* involves the linkage of a chemical group to the terminal amino or carboxyl groups or to reactive groups in the side chains of internal residues; in some cases, these modifications are reversible. *Processing* involves the removal of peptide segments and generally is irreversible.

Acetylation, the addition of an acetyl group (CH_3CO) to the amino group of the N-terminal residue is the most common form of chemical modification, involving an estimated 80 percent of all proteins:

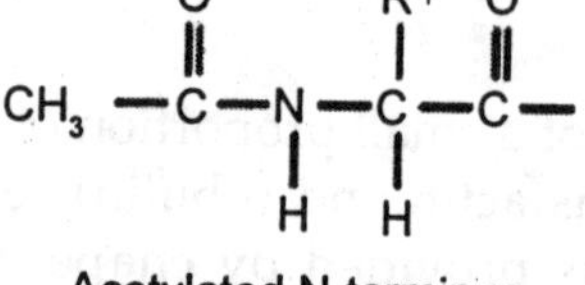

Acetylated N-terminus

This modification may play an important role in controlling the life span of proteins within cells, as nonacetylated proteins are rapidly degraded by intracellular proteases. As discussed later, residues at or near the termini of some membrane proteins are chemically modified by addition of long lipidlike groups. Attachment of these hydrophobic "tails," which function to anchor proteins to the lipid bilayer, constitute one way that cells restrict certain proteins to membranes.

The internal residues in proteins can be modified by attachment of a variety of chemical groups to their side chains. The most important modification is *phosphorylation* of serine, threonine, and tyrosine residues. We will encounter numerous examples of proteins whose activity is regulated by reversible phosphorylation and dephosphorylation. The side chains of asparagine, serine, and threonine are sites for *glycosylation,* the attachment of linear and branched carbohydrate chains. Many secreted proteins and membrane proteins contain glycosylated residues. Various less common modifications are found in a limited number of proteins.

Unlike chemical modification of residues, which often is reversible, processing of some proteins causes irreversible changes that alter their activity. In the most common form of processing, residues are removed from the C- or N-terminus of a polypeptide by cleavage of the peptide bond in a reaction catalyzed by proteases. Proteolytic cleavage is a common mechanism of activation or inactivation, especially of enzymes involved in blood coagulation or digestion. As discussed later, the activity of certain digestive enzymes is controlled by this mechanism. Proteolysis also generates active peptide hormones, such as EGF mentioned earlier and insulin, from larger precursor polypeptides.

An unusual type of processing, termed *protein selfsplicing,* occurs in bacteria and primitive eukaryotes. *Splicing* refers to a process analogous to editing film: an internal segment of polypeptide, an intein, is removed and the ends of the polypeptide are rejoined. Unlike proteolytic processing, protein self-splicing is an autocatalytic process, which proceeds by itself without the involvement of enzymes. The excised peptide appears to eliminate itself from the protein by a mechanism similar to that used in processing of RNA molecules. In vertebrate cells, processing of some proteins involves self-cleavage, but the subsequent ligation step is absent. One such protein is Hedgehog, which is critical to a number of developmental processes.

CELLS DEGRADE PROTEINS VIA SEVERAL PATHWAYS

Cells have both extracellular and intracellular pathways for degrading proteins. The major extracellular pathway is the system of *digestive proteases,* which break down ingested proteins to polypeptides in the intestinal tract. These include *endoproteases* such as trypsin and chymotrypsin, which cleave the protein backbone adjacent to basic and aromatic residues; *exopeptidases,* which sequentially remove residues from the N-terminus (aminopeptidases) or C-terminus (carboxypeptidases) of proteins; and *peptidases,* which split oligopeptides into di- and tripeptides and individual amino acids. These small molecules then are transported across the intestinal lining into the bloodstream.

The life span of intracellular proteins varies from as short as a few minutes for mitotic cyclins, which help regulate passage through mitosis, to as long as the age of an organism for proteins in the lens of the eye. Cells have several intracellular proteolytic pathways for degrading misfolded or denatured proteins, normal proteins whose concentration must be decreased, and foreign proteins taken up by the cell. One major intracellular pathway involves degradation by enzymes within lysosomes, membrane-limited organelles whose interior is acidic. Distinct from the lysosomal pathway are cytosolic mechanisms for degrading proteins. The best-understood pathway, the ubiquitin-mediated pathway, involves two steps: addition of a chain of ubiquitin molecules to an internal lysine side chain of a target protein and proteolysis of the ubiquitinated protein by a proteasome, a large, cylindrical multisubunit complex. The numerous proteasomes present in the cell cytosol proteolytically cleave ubiquitin-tagged proteins in an ATP-dependent process that yields peptides and intact ubiquitin molecules.

To be targeted for degradation by the ubiquitinmediated pathway, a protein must contain a structure that is recognized by a ubiquitinating enzyme complex. Different conjugating enzymes recognize different degradation signals in target proteins. For example, the internal sequence Arg-X-X-Leu-Gly-X-Ile-Gly-Asx in mitotic cyclin is recognized by the ubiquitin-conjugating enzyme E1. Internal sequences enriched in proline, glutamic acid, serine, and threonine (PEST sequences) are recognized by other enzymes. The life span of many cytosolic proteins is correlated with the identity of the N-terminal residue, suggesting that certain residues at the N-terminus favour rapid ubiquitination. For example, short-lived proteins that are degraded within 3 minutes in vivo commonly have Arg, Lys, Phe, Leu, or Trp at their N-terminus. In contrast, a stabilizing amino acid such as Cys, Ala, Ser; Thr, Gly, Val, or Met is present at the N-terminus in long-lived proteins that resist proteolytic attack for more than 30 hours. As explained all newly synthesized proteins have methionine, a stabilizing amino acid, at the N-terminus. Thus subsequent enzymatic alteration that generates one of the destabilizing amino acids at the N-terminus is necessary to target a protein for degradation.

ABERRANTLY FOLDED PROTEINS ARE IMPLICATED IN SLOWLY DEVELOPING DISEASES

As noted earlier, each protein species normally folds into a single, energetically favorable conformation that is specified by its amino acid sequence. Recent evidence suggests, however, that a protein may fold into an alternative three-dimensional structure for reasons that have not yet been identified. Such "misfolding" not only leads to a loss of the normal function of a protein but also marks it for proteolytic degradation. The subsequent accumulation of proteolytic fragments contributes to certain degenerative diseases characterized by the presence of insoluble protein plaques in various organs including the liver and brain.

Alzheimer's disease, for example, is marked by formation of plaques and tangles in a deteriorating brain. The filaments composing these structures are derived from proteolytic products of abundant natural proteins such as amyloid precursor protein, a transmembrane protein, and Tau, a microtubule-binding protein. Plaques in other organs are formed from proteolytic fragments of natural proteins such as gelsolin, an actin-binding protein, and serum albumin, a blood protein. The polypeptide fragments liberated by proteolysis polymerize into very stable filaments. A degeneration of the brain, similar to that seen in Alzheimer's disease, is thought to be caused by *prions*, an infectious protein agent derived by proteolysis and re-folding of a normal brain protein.

FUNCTIONAL DESIGN OF PROTEINS

A key concept in biology is that form and function are inseparable. This concept applies equally well to protein design as to other levels of biological organization (e.g., the morphology of cells and the organization of tissues). In fact, we can often guess how a protein works by looking at its structure. Perhaps the best way to illustrate this is by examining a few protein structures. For instance, a barrel-like nuclear pore, a complex of several proteins, sits in the nuclear membrane and acts as a channel through which molecules travel in or out of the nucleus. In the cavity of a different barrel-like structure, the GroEL/ES chaperonin, protein folding takes place. Some proteins have grooves in their surface, which are logical binding sites for a variety of molecules, especially rod-shaped or filamentous ones. An example is reverse transcriptase, which copies RNA into DNA; this enzyme has a groove on one side through which RNA slides along the surface of the protein. Topoisomerase II, a DNA-binding enzyme, is an articulated enzyme that opens and closes at both ends like locks in a canal. A delight in studying protein structure is uncovering the simple but ingenious ways that nature has built each protein to perform a particular function.

In this section, we examine several features of proteins that are critical to their biological activity and the regulation of that activity, focusing on antibodies, enzymes, and membrane proteins as examples. The functioning of many proteins involves some change in their conformation induced by binding of a specific molecule, change in the environment, or chemical modification. As numerous examples in later chapters will illustrate, such induced conformational changes can make proteins into switches and machines. The changes in conformation can be enormous, as seen in proteins like topoisomerase, an enzyme that moves DNA strands across one another, or myosin, a motor protein that moves along actin filaments.

PROTEINS ARE DESIGNED TO BIND A WIDE RANGE OF MOLECULES

The function of nearly all proteins depends on their ability to bind other molecules, or ligands, with a high degree of specificity. As catalysts of chemical reactions, enzymes must first bind tightly and specifically to their target molecules, called substrates, which may be a small molecule (e.g., glucose) or a macromolecule. The many different types of hormone receptors on the surface of cells also display a high degree of sensitivity and discrimination for their ligands, which generally are present at low concentrations in blood. These receptors, essential to signaling between cells.

Two properties of a protein characterize its interaction with ligands. *Affinity* refers to the strength of binding between a protein and ligand; the equilibrium constant K_{eq} or the dissociation constant K_D for binding is a measure of affinity. *Specificity* refers to the ability of a protein to bind one molecule in preference to other molecules. Both properties depend on the structure of the ligand-binding site on a protein, which is designed to fit its partner like a mold. For high-affinity and highly specific interactions to occur, the shape and chemical surface of the binding site must be complementary to the ligand molecule. To illustrate this critical concept, we consider how an antibody binds an antigen and how an enzyme catalyzes a chemical reaction.

ANTIBODIES EXHIBIT PRECISE LIGAND-BINDING SPECIFICITY

The capacity of proteins to distinguish different molecules is highly developed in blood proteins called antibodies. Animals produce antibodies in response to the invasion of an infectious agent (e.g., a bacterium or a virus) or after exposure to certain foreign substances (e.g., proteins or polysaccharides in pollens). The antibody-inducing agent is called an antigen. The presence of antigen causes an organism to make a large quantity of different antibody proteins, each of which may bind to a slightly different region of the antigen. The constellation of antibodies induced by a given antigen may differ from one member of a species to another.

All antibodies belong to a family of proteins called immunoglobulins. These Y-shaped molecules are formed from two types of polypeptides: heavy chains and light chains. The heavy chains run the length of the molecule; their C-terminal regions pair to form a stem. Visually we can distinguish three globular domains: two identical domains corresponding to each arm and the third composing the stem. Each arm of the antibody molecule contains a single light chain linked to a heavy chain by disulfide bonds. The N-terminal regions of both heavy and light chains lie at the tip of each arm and are distinguished by highly variable amino acid sequences. The remaining portions of the sequences in both chains are constant (i.e., nearly identical) among antibodies with different specificities. The arms are the business end of an antibody molecule, since an antigen-binding site lies at the end of each arm. Because of its dimeric structure, each antibody molecule can bind two identical antigen molecules. X-ray crystallographic analysis of antigen-antibody complexes has revealed that the antigenic specificity of an antibody is dependent on three highly variable regions, called *complementarity-determining regions (CDRs),* near the end of each arm. These regions form the antigen-binding site, which physically matches the antigen like a glove.

Most large antigens have multiple different sites, called epitopes (or antigenic determinants) that can induce production of specific antibodies; each type of antibody binds to its own inducing epitope. For example, lysozyme, an enzyme that degrades the carbohydrate coat of bacteria, induces several different antibodies, each of which binds to a particular epitope on the lysozyme molecule. Although the different epitopes on lysozyme differ greatly in their chemical properties, the interaction between lysozyme and antibody is complementary in all cases; that is, the surface of the antibody's antigen-binding site fits into that of the corresponding epitope as if they were molded together. The intimate contact between these two surfaces, stabilized by numerous noncovalent bonds, is responsible for the exquisite binding specificity exhibited by an antibody. Antibodies, for instance, can distinguish between the cells of individual members of a species and in some cases can distinguish between proteins that differ by only a single amino acid. Because of their specificity and the ease with which they can be produced, antibodies are critical reagents in many experiments discussed in the following chapters.

ENZYMES ARE HIGHLY EFFICIENT AND SPECIFIC CATALYSTS

Almost every chemical reaction in a cell is catalyzed by a class of proteins called enzymes. As discussed catalysts increase the rates of reactions that are already energetically favorable by lowering the activation energy. In the test tube, catalysts such as charcoal and platinum facilitate reactions but often at high temperatures, at extremes of high or low pH, or in organic solvents. In contrast to these harsh conditions, enzymes must catalyze chemical reactions in the mild conditions of a cell: 37 °C, pH 6.5–7.5, and aqueous solvents. As we just discussed, all antibodies belong to the immunoglobulin family of proteins and have a similar structure. Enzymes, however, are a structurally diverse group of proteins that have evolved through unrelated and highly divergent mechanisms.

The ability of enzymes to function as catalysts under conditions where nonbiological catalysts would be ineffectual is exemplified by two striking properties: their enormous reaction rates and their specificity. Quite often, the rate of an enzymatically catalyzed reaction is 10^6–10^{12} times that of an uncatalyzed reaction under otherwise similar conditions. The specificity of an enzyme denotes its ability to act selectively on one substance or a small number of chemically similar substances, the enzyme's substrates. Like antibody specificity, enzyme specificity depends on a close fit between substrate molecules and their binding sites on an enzyme. An example of specificity is provided by the enzymes that act on amino acids. As noted amino acids can exist as two stereoisomers, designated L and D, although only L isomers normally are found in biological systems. Not surprisingly, enzyme-catalyzed reactions involving L-amino acids occur much more rapidly than do those involving D-amino acids, even though both stereoisomers of a given amino acid are the same size and possess the same R groups.

The number of different types of chemical reactions that occur in any one cell is very large: an animal cell, for example, normally contains 1000–4000 different types of enzymes, each of which catalyzes a single chemical reaction or set of closely related reactions. Certain enzymes are found in the majority of cells because they catalyze synthesis of common cellular products (e.g., proteins, nucleic acids, and phospholipids) or are involved in the production of energy by the conversion of glucose and oxygen to carbon dioxide and water. Other enzymes are present only in a particular type of cell (e.g., a liver cell or a nerve cell) because they catalyze some chemical reaction unique to that cell type. Although most enzymes are located within cells, some are secreted and function in the blood, lumen of the digestive tract, or other extracellular space. Some microbial enzymes are secreted from and are active outside the organism.

AN ENZYME'S ACTIVE SITE BINDS SUBSTRATES AND CARRIES OUT CATALYSIS

Certain amino acid side chains of an enzyme are important in determining its specificity and its ability to accelerate a chemical reaction. In the native conformation of an enzyme, these side chains are brought into proximity, forming the active site. Active sites thus consist of two functionally important regions: one that recognizes and binds the substrate (or substrates), and one that catalyzes the reaction once the substrate has been bound. In some enzymes, the catalytic site is part of the substrate-binding site; in others, the two sites are structurally as well as functionally distinct. The amino acids that make up the active site do not need to be adjacent in the linear polypeptide sequence; rather, folding of the molecule results in juxtaposition of these amino acids, forming a space in which the substrate sits.

To illustrate how the active site binds a specific substrate and then promotes a chemical change in the bound substrate, we examine the action of *cAMP-dependent protein kinase (cAPK)*. This enzyme and other protein kinases, which add a phosphate group to serine, threonine, or tyrosine residues in proteins, are critical for regulating the activity of many cellular proteins. Because the structure of the active site and mechanism of phosphorylation are very similar in all kinases, cAPK can serve as a general model for this important class of enzymes.

As discussed later, an unusual nucleotide called *cAMP* induces dissociation of the inactive tetrameric form of cAPK, releasing two catalytic subunits. To aid in understanding the mechanism of binding and catalysis, we focus here on the 260-residue "kinase core" of each catalytic subunit. The kinase core, which is largely conserved in all protein kinases, is responsible for the binding of ATP and a target peptide, followed by transfer of a phosphate group from ATP to a serine, threonine, or tyrosine in the peptide. The kinase core consists of a large and small domain with an intervening deep cleft; the active site comprises residues located in both domains.

SUBSTRATE BINDING BY PROTEIN KINASES

The small domain of the kinase core binds ATP, while the large domain binds the target peptide. The structure of the ATP-binding site complements the structure of the nucleotide substrate. The adenine ring of ATP sits snugly at the base of the cleft, which is characterized by a highly conserved sequence, Gly-X-Gly-X-X-Gly. This triad of glycine residues, the "glycine lid," is part of a strand-loop-strand motif that closes over the adenine of ATP and holds it in position. The adenine ring sits in a hydrophobic pocket and is positioned by hydrogen bonds and van der Waals attractions with the glycine residues and backbone amide groups. Two invariant residues, lysine at position 72 and aspartic acid at position 184, stabilize the phosphate groups, which protrude from the nucleotide-binding cleft. Lys-72 bridges to the α and β phosphates of ATP, while the γ-phosphate group is chelated by a Mg^{2+} ion bound to Asp-184.

ATP is a common substrate for all protein kinases, but the sequence of the target peptide varies among different kinases. The peptide sequence recognized by cAPK is Arg-Arg-X-Ser-Y, where X is any amino acid and Y is a hydrophobic amino acid. The portion of the polypeptide chain containing the target serine, threonine, or tyrosine is bound to a shallow groove in the large domain of the kinase core. The peptide specificity of cAPK is conferred by several glutamic acid residues in the large domain, which bind the two arginine residues in the target peptide. Different residues determine the specificity of other protein kinases.

PHOSPHORYL TRANSFER BY CAPK

Catalysis takes place in two stages. First, a bond forms between the serine and phosphate group, yielding a pentavalent phospmhate transition state. Second, the phosphodiester bond between the β and γ phosphates is broken, yielding ADP and the phosphorylated peptide. Because the phosphate-serine bond is formed on the opposite side of the phosphodiester bond from the β phosphate, this process is called an *in-line mechanism of phosphoryl transfer*. The products, ADP and phosphorylated peptide, are then released from the active site.

During catalysis by cAPK, two "catalytic residues" appear to participate in formation of the transition state. Asp-166 is thought to remove a proton from the serine hydroxyl group in the target peptide, while Lys-168 neutralizes the negative charge of the γ phosphate. Then the electrons of the deprotonated serine hydroxyl group are thought to form a bond to the γ phosphorus atom, yielding the pentavalent transition-state intermediate. The newly created phosphoserine is repelled from the β phosphate of ADP and the catalytic base. The products induce a conformational change in the enzyme, described below, that permits them to diffuse from the active site.

Interactions between residues in the active site of an enzyme and the substrates help stabilize the transition state, thereby allowing more time for the rearrangement of bonds needed to form the products. As explained the activation energy is the energy required for formation of the transition state. An enzyme, by virtue of its three-dimensional binding site, reduces the activation energy of a reaction compared with an uncatalyzed reaction involving the same reactants. The ability to bind transition-state intermediates is the one property that distinguishes enzymes from other proteins. If a protein cannot bind a transition-state intermediate, then it cannot catalyze a reaction.

CONFORMATIONAL CHANGES INDUCED BY SUBSTRATE BINDING TO CAPK

The catalytic subunit of cAPK exists in an "open" and "closed" conformation. In the open position, the large and small domains of the kinase core are separated enough that substrate molecules can bind. Once the active site is occupied by substrate, the domains move together into the closed position. This change in tertiary structure, an example of *induced fit,* brings the bound target peptide close enough to the terminal phosphate group of the bound ATP that phosphoryl transfer can occur. After the phosphorylation reaction is completed, the presence of the products causes the domains to rotate to the open position, from which the products are released.

The rotation from the open to closed position also causes movement of the short glycine-rich sequence over the ATP-binding cleft in the active site. This small finger of the polypeptide chain, the glycine lid, controls the entry of ATP and release of ADP at the active site. In the open position, ATP can enter and bind to the active site cleft. In the closed position, the glycine-rich sequence moves over the nucleotide and acts as a lid that prevents ATP from leaving. Following phosphoryl transfer, the glycine lid must rotate back to the open position before ADP can be released. Kinetic measurements show that the rate of ADP release is 20-fold slower than that of phosphoryl transfer, reflecting the influence of the glycine lid in cAPK. Mutations in the glycine lid that inhibit its flexibility slow catalysis by cAPK even further. Besides trapping ATP in the binding pocket, the glycine lid prevents water from entering the active site. Water would inhibit the reaction by dampening the charge delocalization steps.

KINETICS OF AN ENZYMATIC REACTION ARE DESCRIBED BY V_{MAX} AND K_M

Enzymatic specificity is usually quantified in relative terms; that is, the reaction with a good substrate may occur, for example, 10,000 times faster than it does with a poor substrate. The catalytic action of an enzyme on a given substrate can be described by two parameters: K_m (the Michaelis constant), which measures the affinity of an enzyme for its substrate, and V_{max}, which measures the maximal velocity of the reaction at saturating substrate concentrations. Equations for K_m and V_{max} are most easily derived by considering the simple reaction

$$\textit{Substrate} \rightleftharpoons \textit{product}$$

in which the rate of product formation v depends on the concentration of substrate, *[S]*, and on the concentration of the enzyme, *[E]*.

For an enzyme with a single catalytic site, shows how the rate of product formation depends on *[S]* when *[E]* is kept constant. At low concentrations of S, the reaction rate is proportional to *[S]*. As *[S]* is increased, the rate does not increase indefinitely in proportion to *[S]*; rather, it eventually reaches a maximum velocity V_{max}. The value of V_{max} is independent of *[S]*, but is proportional to *[E]* and to the catalytic constant k_{cat}, which is an intrinsic property of the individual enzyme. Halving *[E]* reduces the reaction rate at all values of *[S]* by half. Both V_{max} and K_m for a particular enzyme and substrate can be determined from experimental curves of reaction velocity versus substrate concentration.

When interpreting kinetic curves bear in mind that all enzyme-catalyzed reactions include at least three steps:

1. The binding of a substrate (S) to an enzyme (E) to form an enzyme-substrate complex (ES),
2. The conversion of ES to the enzyme-product complex (EP),
3. The release of the product (P) from EP, to yield free P:

$$E + S \underset{}{\overset{binding}{\rightleftharpoons}} ES \xrightarrow{catalysis} EP \xrightarrow{release} P + E$$

In the simplest case, when the release of P is very rapid, we can simplify the reaction equation as follows:

$$E + S \underset{K_2}{\overset{K_1}{\rightleftharpoons}} ES \xrightarrow{K_{cat}} E + P$$

In this case, the rate of product formation v is equal to k_{cat} × *[ES]*. Starting from this relationship, we can derive the *Michaelis-Menten equation*

$$\upsilon = V_{\max} \frac{[S]}{[S]+[K_m]}$$

where K_m, the Michaelis constant, is defined as $(k_2+k_{cat})/k_1$. This equation fits the curves.

The slowest step in most enzymatic reactions is conversion of the enzyme-substrate complex ES to the free enzyme E and product P. In such cases, k_{cat} is much less than k_2, so that

$$K_m \cong \frac{k_2}{k_1} = K_d$$

Where K_d is the dissociation constant for binding of S to E. Thus the parameter K_m describes the affinity of an enzyme for its substrate. The smaller the value of K_m, the more avidly the enzyme can bind the substrate from a dilute solution and the smaller the concentration of substrate needed to reach half-maximal velocity. The concentrations of the various small molecules in a cell vary widely, as do the K_m values for the different enzymes that act on them. Generally, the intracellular concentration of a substrate is approximately the same as or greater than the K_m value of the enzyme to which it binds.

MANY PROTEINS CONTAIN TIGHTLY BOUND PROSTHETIC GROUPS

The native conformation and activities of some proteins require the presence of a prosthetic group, a small nonpeptide molecule or metal that binds tightly to a protein, keeping the protein in a fixed conformation and participating in binding ligands. For example, each of the four subunits of hemoglobin binds and enfolds a prosthetic group called *heme,* which consists of an iron atom held in a cage by protoporphyrin:

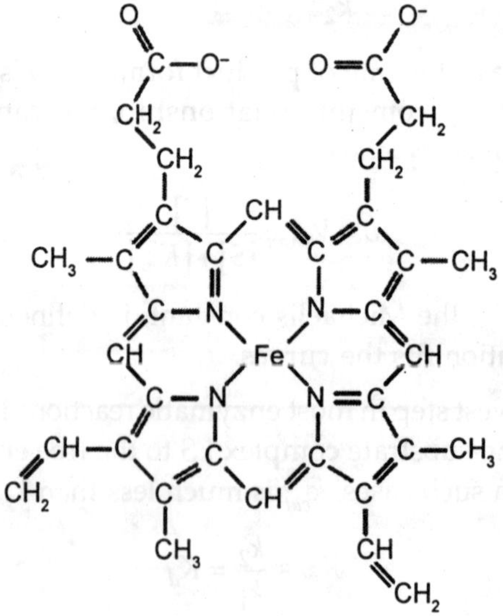

The heme groups are the oxygen-binding components of hemoglobin. Heme is also present in the cytochromes of the electron-transport chain; in this case, it functions to bind electrons. Other electron-transport proteins employ sulfur or flavin as prosthetic groups. In addition to acting as carriers of oxygen or electrons, prosthetic groups can act as antennae. For example, proteins involved in vision or photosynthesis contain retinal or chlorophyll, which absorb energy from sunlight. Prosthetic groups can be linked to proteins noncovalently, as in hemoglobin, or covalently, as in cytochrome.

The activity of numerous enzymes also depends on the presence a prosthetic group, commonly referred to as a coenzyme. Many coenzymes act to lower the activation energy of biochemical reactions by forming a covalent intermediate with a substrate. For instance, the enzyme that converts the amino acid histidine into histamine (a potent dilator of small blood vessels) requires the coenzyme *pyridoxal phosphate.* In this reaction, a covalent bond first forms between histidine and the enzyme-bound pyridoxal phosphate, forming a Schiff base intermediate. Rearrangement of the bonds in this intermediate yields carbon dioxide, which is released, and a second intermediate. This is then hydrolyzed, producing the product, histamine, and regenerating the coenzyme, pyridoxal phosphate.

A VARIETY OF REGULATORY MECHANISMS CONTROL PROTEIN FUNCTION

Most reactions in cells do not occur independently of one another or at a constant rate. Instead, the catalytic activity of enzymes is so regulated that the amount of reaction product is just sufficient to meet the needs of the cell. As a result, the steady-state concentrations of substrates and products will vary depending on cellular conditions. The flow of material in an enzymatic pathway is controlled by several mechanisms, some of which also regulate the functions of nonenzymatic proteins.

One of the most important mechanisms for regulating protein function entails allosteric transitions, changes in the tertiary and/or quaternary structure of a protein induced by binding of a small molecule, which may be an activator, inhibitor, or substrate. Allosteric regulation is particularly prevalent in multimeric (multisubunit) enzymes. Some multimeric enzymes are composed of identical subunits, each containing an active site and, often, a distinct regulatory site. Other enzymes comprise structurally different subunits; in these, active sites and regulatory sites may be located on different subunits.

ALLOSTERIC RELEASE OF CATALYTIC SUBUNITS

As mentioned previously, cAMP-dependent protein kinase (cAPK) exists as an inactive tetrameric protein composed of two catalytic subunits and two regulatory subunits. Each regulatory subunit contains a *pseudosubstrate* sequence that binds to the active site in a catalytic subunit. By blocking substrate binding, the regulatory subunit inhibits the activity of the catalytic subunit. Binding of the allosteric effector molecule cyclic AMP (cAMP) to the regulatory subunit induces a conformational change in the pseudosubstrate sequence so that it no longer can bind the catalytic subunit. Thus the inactive tetramer dissociates into two monomeric active catalytic subunits and a dimeric regulatory subunit. As discussed binding of various hormones to cell-surface receptors induces a rise in the intracellular concentration of cAMP, leading to activation of cAPK. Once the signaling ceases and the cAMP level decreases, the activity of cAPK is turned off by reassembly of the inactive tetramer.

ALLOSTERIC TRANSITION BETWEEN ACTIVE AND INACTIVE STATES

Many multimeric enzymes undergo allosteric transitions that alter the relationship of the subunits to one another but do not cause dissociation as in cAPK. A wellunderstood enzyme illustrating this mechanism is aspartate transcarbamoylase (ATCase). This bacterial enzyme catalyzes the first step in the pyrimidine biosynthetic pathway:

$$\textit{Aspartate + carbamoyl phosphate} \xrightarrow{\textit{ATCcase}} \textit{N - carbamoylaspartate}$$

ATCase, which is composed of six catalytic subunits and six regulatory subunits, exists in an active R state and inactive T state. The equilibrium between these states is shifted toward the inactive T state by binding of cytidine triphosphate (CTP), an end product of the pyrimidine pathway, to the regulatory subunits. Thus CTP is an allosteric inhibitor of ATCase. The CTP-induced allosteric transition in ATCase is an example of *feedback inhibition,* whereby an enzyme that catalyzes an early reaction in a multistep pathway is inhibited by an ultimate product of the pathway. Clearly, this type of regulation prevents accumulation of pyrimidines in excess of what the cell needs for DNA synthesis.

The mechanism of feedback inhibition helps regulate most biosynthetic pathways; that is, the final product of the pathway inhibits the enzyme that catalyzes the first step, thus preventing both production of the intermediate products and unnecessary metabolic activity. Feedback inhibition of enzyme function is reversible. If the concentration of free feedback inhibitor (e.g., CTP) falls, the bound inhibitor dissociates from the regulated enzyme, which then reverts to its active conformation. The binding of a feedback inhibitor to an enzyme and its subsequent release can be described by the equilibrium binding constant K_i, which is similar to the Michaelis constant K_m used to describe substrate binding.

COOPERATIVE BINDING OF LIGANDS

In many cases, especially when a protein binds several molecules of one ligand, the binding is graded; that is, binding of one ligand molecule affects the binding of subsequent ligand molecules. Such *cooperative allostery,* or cooperative binding, permits many multisubunit proteins to respond more efficiently to small changes in ligand concentration than would otherwise be possible. In positive cooperativity, sequential binding is enhanced; in negative cooperativity, sequential binding is inhibited.

Hemoglobin presents a classic example of positive cooperative binding. Each of the four subunits in hemoglobin can bind one oxygen molecule. Binding of oxygen to one subunit induces a local conformational change whose effect spreads to the other subunits, lowering the K_m for binding of additional oxygen molecules.

Many multimeric enzymes, including aspartate transcarbamoylase (ATCase), also exhibit cooperative binding of substrate. For reactions catalyzed by such enzymes, a plot of reaction velocity versus substrate concentration yields a sigmoidal curve rather than the hyperbolic curve characteristic of enzymes with typical Michaelis-Menten kinetics. As a result of cooperative substrate binding, the maximal enzyme activity (V_{max}) is achieved over a narrow range of substrate concentration. Other multimeric enzymes exhibit cooperative binding of an allosteric inhibitor. Because of cooperative allostery, a quite small change in ligand concentration can effectively turn an enzyme on or off.

Cyclic Protein Phosphorylation and Dephosphorylation

As noted earlier, one of the most common mechanisms for regulating protein activity is the addition and removal of phosphate groups from serine, threonine, or tyrosine residues. Protein kinases catalyze phosphorylation, and phosphatases catalyze dephosphorylation. Although both reactions are essentially irreversible, the counteracting activities of kinases and phosphatases provide cells with a "switch" that can turn on or turn off the function of various proteins. Phosphorylation changes a protein's charge and generally leads to a conformational change; these effects can significantly alter ligand binding by a protein and its activity.

Nearly 3 percent of all yeast proteins are protein kinases or phosphatases, reflecting the importance of phosphorylation and dephosphorylation reactions even in simple cells. These enzymes target all classes of proteins including structural proteins, enzymes, membrane channels, and signaling molecules. In later chapters, we will encounter many examples of important cellular functions that are controlled by phosphorylation and dephosphorylation of specific proteins.

PROTEOLYTIC ACTIVATION

The regulatory mechanisms discussed so far can act like switches, turning proteins on and off. Regulation of some enzymes involves the irreversible activation of an inactive form, commonly at the site where the active enzyme is needed. Good examples of such enzymes are the digestive proteases trypsin and chymotrypsin, which are synthesized in the pancreas and secreted into the small intestine as inactive precursors, or *zymogens,* called *trypsinogen* and *chymotrypsinogen,* respectively. In the protease-rich environment of the small intestine, the zymogens are converted to the active enzymes, which then begin to hydrolyze the peptide bonds of ingested proteins. Two irreversible proteolytic cleavages of chymotrypsinogen yield active chymotrypsin. The delay in activation of these proteases until they reach the intestine prevents them from digesting the pancreatic tissue in which they are made.

OTHER REGULATORY MECHANISMS

The activities of enzymes are extensively regulated in order that the numerous enzymes in a cell can work together harmoniously. All metabolic pathways are closely controlled at all times. Synthetic reactions occur when the products of these reactions are needed; degradative reactions occur when molecules must be broken down. All the regulatory mechanisms described above affect enzymes locally at their site of action.

Regulation of cellular processes, however, involves more than simply turning enzymes on and off. Some regulation is accomplished by keeping enzymes in compartments where the delivery of substrate or exit of product is controlled. In many cases, the compartments are organelles, such as the mitochondria, nuclei, or lysosomes. Compartmentation permits competing reactions to occur simultaneously in different parts of a cell. In addition to compartmentation, cellular processes are regulated by enzyme synthesis and destruction. Often enzymes are synthesized at low rates when the cell has no need for their activities; however, upon increased demands by the cell (for instance, appearance of substrate), new enzyme is synthesized. Later, the pool of enzyme is lowered when levels of substrate decrease or the cell becomes inactive.

MEMBRANE PROTEINS

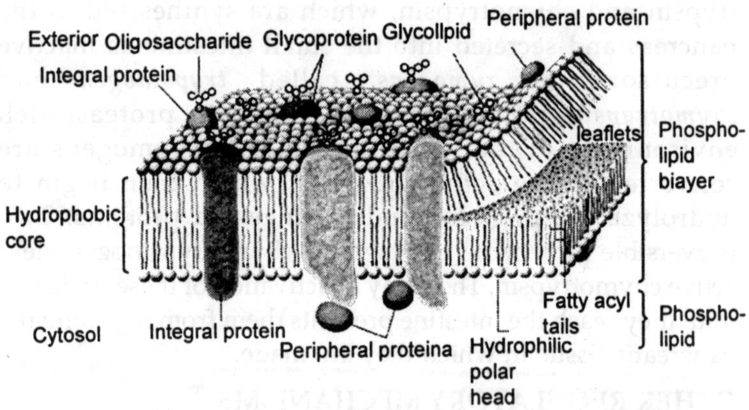

Fig. Schematic diagram of typical membrane proteins in a biological membrane.

As we've seen, all antibodies have a similar structure and function; enzymes are structurally varied, but all have a catalytic function. In contrast, although all membrane proteins are located at the membrane, they otherwise are both structurally and functionally diverse. As we noted and discuss every biological membrane has the same basic phospholipid bilayer structure. Associated with each membrane is a set of membrane proteins that enables the membrane to carry out its distinctive activities. The complement of proteins attached to a membrane varies depending on cell type and subcellular location.

Some proteins are bound only to the membrane surface, whereas others have one region buried within the membrane and domains on one or both sides of it. Protein domains on the extracellular membrane surface are generally involved in cell-cell signaling or interactions. Domains within the membrane, particularly those that form channels and pores, move molecules across the membrane. Domains lying along the cytosolic face of the membrane have a wide range of functions, from anchoring cytoskeletal proteins to the membrane to triggering intracellular signaling pathways. In many cases, the function of a membrane protein and the topology of its polypeptide chain in the membrane can be predicted based on its homology with another, well-characterized protein. In this section, we examine the characteristic structural features of membrane proteins and some of their basic functions. More complete characterization of the structure and function of various types of membrane proteins is presented in several later chapters.

PROTEINS INTERACT WITH MEMBRANES IN DIFFERENT WAYS

Membrane proteins can be classified into two broad categories—integral (intrinsic) and peripheral (extrinsic)—based on the nature of the membrane-protein interactions. Most biomembranes contain both types of membrane proteins.

Integral membrane proteins, also called *intrinsic proteins,* have one or more segments that are embedded in the phospholipid bilayer. Most integral proteins contain residues with hydrophobic side chains that interact with fatty acyl groups of the membrane phospholipids, thus anchoring the protein to the membrane. Most integral proteins span the entire phospholipid bilayer. These *transmembrane* proteins contain one or more membrane-spanning domains as well as domains, from four to several hundred residues long, extending into the aqueous medium on each side of the bilayer. In all the transmembrane proteins examined to date, the membrane-spanning domains are α helices or multiple β strands. In contrast, some integral proteins are anchored to one of the membrane leaflets by covalently bound fatty acids, as discussed later. In these proteins, the bound fatty acid is embedded in the membrane, but the polypeptide chain does not enter the phospholipid bilayer.

Peripheral membrane proteins, or extrinsic proteins, do not interact with the hydrophobic core of the phospholipid bilayer. Instead they are usually bound to the membrane indirectly by interactions with integral membrane proteins or directly by interactions with lipid polar head groups. Peripheral proteins localized to the cytosolic face of the plasma membrane include the cytoskeletal proteins spectrin and actin in erythrocytes and the enzyme protein kinase C. This enzyme shuttles between the cytosol and the cytosolic face of the plasma membrane and plays a role in signal transduction. Other peripheral proteins, including certain proteins of the extracellular matrix, are localized to the outer (exoplasmic) surface of the plasma membrane.

HYDROPHOBIC α HELICES IN TRANSMEMBRANE PROTEINS ARE EMBEDDED IN THE BILAYER

Integral proteins containing membrane-spanning α-helical domains are embedded in membranes by hydrophobic interactions with the lipid interior of the bilayer and probably also by ionic interactions with the polar head groups of the phospholipids. *Glycophorin,* a major erythrocyte membrane protein, exhibits both types of interaction. Glycophorin contains a membrane-embedded α helix composed entirely of hydrophobic (or uncharged) amino acids. The predicted length of this α helix (3.75 nm) is just sufficient to span the hydrocarbon core of a phospholipid bilayer. The hydrophobic side chains form van der Waals interactions with the fatty acyl chains and shield the polar carbonyl (C=O) and imino (NH) groups of the peptide bond, which are all hydrogen-bonded to one another. This hydrophobic helix is prevented from slipping across the membrane by a flanking set of positively charged amino acids (lysine and arginine) that are thought to interact with negatively charged phospholipid head groups. In glycophorin, most of these charged residues lie adjacent to the cytosolic leaflet.

MANY INTEGRAL PROTEINS CONTAIN MULTIPLE TRANSMEMBRANE α HELICES

Although glycophorin as a monomer with a single α helix spanning the bilayer, this protein is present in erythrocyte membranes as a dimer of two identical polypeptide chains. The two membrane-spanning α helices of glycophorin are thought to form a coiled-coil structure stabilized by specific interactions between the amino acid side chains at the interface of the two helices. It is now known that many other transmembrane proteins contain two or more membrane-spanning α helices. For instance, the *bacterial photosynthetic reaction centre (PRC)* comprises four subunits and several prosthetic groups, including four chlorophyll molecules. In this complex protein, three of the four subunits span the membrane; two of these subunits (L and M) each contain five membrane-spanning α helices.

A large and important family of integral proteins is defined by the presence of seven membrane-spanning α helices. More than 150 such "seven-spanning" membrane proteins have been identified. This class of integral proteins is typified by *bacteriorhodopsin,* a protein found in a photosynthetic bacterium. Absorption of light by the retinal group attached to bacteriorhodopsin causes a conformational change in the protein that results in pumping of protons from the cytosol across the bacterial membrane to the extracellular space. The proton concentration gradient thus generated across the membrane is used to synthesize ATP. Both the overall arrangement of the seven α helices in bacteriorhodopsin and the identity of most of the amino acids can be resolved by computer analysis of micrographs of two-dimensional crystals of the membrane-embedded protein taken at various angles to the electron beam.

Other seven-spanning membrane proteins include the opsins (eye proteins that absorb light), cell-surface receptors for many hormones, and receptors for odorous molecules. Amino acid sequence analysis of these proteins has shown that no amino acids are found in the same position in all of them, and only a few residues are conserved in even a substantial number of them. Nonetheless, each of these proteins contains seven stretches of hydrophobic amino acids long enough (>22 amino acids) to span the phospholipid bilayer. Though direct evidence is lacking, it is thought that all of these proteins adopt a conformation in the membrane similar to that of bacteriorhodopsin. This is one of several examples of how investigators can predict the orientation of proteins in a membrane from the amino acid sequence alone.

MULTIPLE β STRANDS IN PORINS FORM MEMBRANE-SPANNING "BARRELS"

The *porins* are a class of transmembrane proteins whose structure differs radically from that of other integral proteins. Several types of porin are found in the outer membrane of gram-negative bacteria such as *E. coli*. The outer membrane protects an intestinal bacterium from harmful agents (e.g., antibiotics, bile salts, and proteases) but permits the uptake and disposal of small hydrophilic molecules including nutrients and waste products. The porins in the outer membrane of an *E. coli* cell provide channels for passage of disaccharides, phosphate, and similar molecules.

The amino acid sequences of porins are predominantly polar and contain no long hydrophobic segments typical of integral proteins with α-helical membrane-spanning domains. X-ray crystallography has revealed that porins are trimers of identical subunits. In each subunit 16 β strands form a barrel-shaped structure with a pore in the centre. As noted earlier, half the amino acid side groups of a β strand point in one direction, and the other half point in the opposite direction. Unlike a typical globular protein, porins have an inside-out arrangement. In a porin monomer, the outward-facing side groups on each of the β strands are hydrophobic and thus can interact with the fatty acyl groups of the membrane lipids or with other porin monomers. The side groups facing the inside of a porin monomer are predominantly hydrophilic; these line the pore through which small water-soluble molecules cross the membrane.

COVALENTLY ATTACHED HYDROCARBON CHAINS ANCHOR SOME PROTEINS TO THE MEMBRANE

In eukaryotic cells, as noted earlier, the polypeptide chain of some integral membrane proteins does not enter the bilayer but rather is anchored in one leaflet by a covalently attached hydrocarbon chain.

Some cell-surface proteins are anchored to the exoplasmic face of the plasma membrane by a complex glycosylated phospholipid that is linked to the C-terminus. A common example of this type of anchor is *glycosylphosphatidylinositol,* which contains two fatty acyl groups, *N*-acetylglucosamine, mannose, and inositol. Several enzymes, including alkaline phosphatase, fall into this class. Various experiments have shown that the phospholipid anchor is both necessary and sufficient for binding these cell-surface proteins to the membrane. For instance, the enzyme phospholipase C cleaves the phosphate-glycerol bond in phospholipids as well as in glycosylphosphatidylinositol anchors, and treatment of cells with phospholipase C releases glycosylphosphatidylinositol-anchored proteins such as Thy-1 protein and alkaline phosphatase from the cell surface.

Some cytosolic proteins are anchored to the cytosolic face of membranes by a hydrocarbon moiety covalently attached to a cysteine near the C-terminus. The most common anchors are prenyl, farnesyl, and geranylgeranyl groups. These proteins undergo a chemical modification involving several steps. First, the anchor moiety forms a thioether bond with the thiol group of a cysteine that is four residues from the C-terminus of the protein. The modified protein then undergoes proteolysis and methylation; these reactions remove the three terminal residues and add a methyl to the new C-terminus. In some cases, fatty acyl palmitate groups form thioester bonds to nearby cysteine residues, providing additional anchors that are thought to reinforce the attachment of the protein to the membrane.

In another group of lipid-anchored cytosolic proteins, a fatty acyl group (e.g., myristate or palmitate) is linked by an amide bond to the N-terminal glycine residue. In these proteins, the N-terminal anchor is necessary for retention at the membrane and may play an important role in a membrane-associated function. For example, v-Src, a mutant form of a cellular tyrosine kinase, is oncogenic and can transform cells only when it retains a myristylated N-terminus.

Some Peripheral Proteins Are Soluble Enzymes That Act on Membrane Components

An important group of peripheral membrane proteins are water-soluble enzymes that associate with the polar head groups of membrane phospholipids. One well-understood group of such enzymes are the *phospholipases,* which hydrolyze various bonds in the head groups of phospholipids. These enzymes have an important role in the degradation of damaged or aged cell membranes.

The mechanism of action of phospholipase A_2 illustrates how such water-soluble enzymes can reversibly interact with membranes and catalyze reactions at the interface of an aqueous solution and lipid surface. When this enzyme is in aqueous solution, its Ca^{2+}-containing active site is buried in a channel lined with hydrophobic amino acids. Binding of the enzyme to a phospholipid bilayer induces a small conformational change that fixes the protein to the phospholipid heads and opens the hydrophobic cleft. As a phospholipid molecule moves from the bilayer into the channel, the enzyme-bound Ca^{2+} binds to the phosphate in the head group and positions the ester bond to be cleaved next to the catalytic site.

PURIFYING, DETECTING, AND CHARACTERIZING PROTEINS

A protein must be purified before its structure and the mechanism of its action can be studied. However, because proteins vary in size, charge, and water solubility, no single method can be used to isolate all proteins. To isolate one particular protein from the estimated 10,000 different proteins in a cell is a daunting task that requires methods both for separating proteins and for detecting the presence of specific proteins.

Any molecule, whether protein, carbohydrate, or nucleic acid, can be separated from other molecules based on large differences in some physical characteristic. Although the sequence of amino acids in a protein uniquely determines its function, the most useful physical characteristic for separation of proteins is *size*, defined as either length or mass. In this section, we briefly outline different techniques for separating proteins based on their size and other properties. These techniques also apply to the separation of nucleic acids and other biomolecules. We then consider general methods for detecting, or *assaying*, specific proteins, including the use of radioactive compounds for tracking biological activity. Finally, we discuss several techniques for characterizing a protein's mass, sequence, and three-dimensional structure.

PROTEINS CAN BE REMOVED FROM MEMBRANES BY DETERGENTS OR HIGH-SALT SOLUTIONS

Because water-soluble globular proteins have many exposed hydrophilic groups, they maintain their native conformation and remain individually suspended in an aqueous medium when separated from cells. In contrast, when transmembrane proteins are separated from membranes, their exposed hydrophobic regions interact, causing the protein molecules to aggregate and precipitate from aqueous solutions. Such proteins can be solubilized by detergents, which have affinity both for hydrophobic groups and for water.

Detergents are amphipathic molecules that disrupt membranes by intercalating into phospholipid bilayers and solubilizing lipids and proteins. The hydrophobic part of a detergent molecule is attracted to hydrocarbons and mingles with them readily; the hydrophilic part is strongly attracted to water. Some detergents are natural products, but most are synthetic molecules developed for cleaning and for dispersing mixtures of oil and water. Ionic detergents, such as sodium deoxycholate and sodium dodecylsulfate (SDS), contain a charged group; nonionic detergents, such as Triton X-100 and octylglucoside, lack a charged group.

At very low concentrations, detergents dissolve in pure water as isolated molecules. As the concentration increases, the molecules begin to form *micelles*. These are small, spherical aggregates in which hydrophilic parts of the molecules face outward and the hydrophobic parts cluster in the centre. The *critical micelle concentration (CMC)* at which micelles form is characteristic of each detergent and is a function of the structures of its hydrophobic and hydrophilic parts.

Ionic detergents bind to the exposed hydrophobic regions of membrane proteins as well as to the hydrophobic core of water-soluble proteins. Because of their charge, these detergents also disrupt ionic and hydrogen bonds. At high concentrations, for example, sodium dodecylsulfate completely denatures proteins by binding to every side chain. *Nonionic detergents* act in different ways at different concentrations. At high concentrations (above the CMC), they solubilize biological membranes by forming mixed micelles of detergent, phospholipid, and integral membrane proteins. At low concentrations (below the CMC), these detergents may bind to the hydrophobic regions of most membrane proteins, making them soluble in aqueous solution. In this case, although mixed micelles are not formed, the solubilized protein will not aggregate during subsequent purification steps.

Most peripheral proteins are bound to specific integral membrane proteins by ionic or other weak interactions. Generally they can be removed from the membrane by solutions of high ionic strength (high salt concentrations), which disrupt ionic bonds, or by chemicals that bind divalent cations such as Mg^{2+}. Unlike integral proteins most peripheral proteins are soluble in aqueous solution and are not solubilized by nonionic detergents.

CENTRIFUGATION CAN SEPARATE PARTICLES AND MOLECULES THAT DIFFER IN MASS OR DENSITY

The first step in a typical protein-purification scheme is centrifugation. The principle behind centrifugation is that two particles in suspension (cells, organelles, or molecules), having different masses or densities will settle to the bottom of a tube at different rates. Remember, *mass* is the weight of a sample (measured in grams), whereas *density* is the ratio of its weight to volume (grams/liter). Proteins vary greatly in mass but not in density. The average density of a protein is 1.37 g/cm^3. Unless a protein has an attached lipid or carbohydrate, its density will not vary by more than 15 percent from this value. Heavier or more dense molecules settle, or sediment, more quickly than lighter or less dense molecules.

A centrifuge speeds sedimentation by subjecting particles in suspension to centrifugal forces as great as 600,000 times the force of gravity g. The centrifugal force is proportional to the rotation rate of the rotor (measured in revolutions per minute, or rpm) and the distance of the tube from the centre of the rotor. Modern ultracentrifuges reach speeds of 60,000 rpm or greater and generate forces sufficient to sediment particles with masses greater than 10,000 daltons (Da). However, small particles with masses of 5 Da or less will not sediment uniformly even at such high rotor speeds.

Centrifugation is used for two basic purposes:

1. As a preparative technique to separate one type of material from others
2. As an analytical technique to measure physical properties (e.g., molecular weight, density, shape, and equilibrium binding constants) of macromolecules. The *sedimentation constant, s,* of a protein equals its velocity in a centrifugal field divided by the centrifugal force. The value of s depends on a protein's density and shape, as well as the density and viscosity of the medium. Because the centrifugal force and the density and viscosity of the medium are all known, the radius and mass of a molecule

can be calculated from measurements of its rate of movement in an analytical ultracentrifuge. The sedimentation constant is commonly expressed in svedbergs (S): 1 S = 10^{-13} seconds.

DIFFERENTIAL CENTRIFUGATION

The most common initial step in protein purification is separation of soluble proteins from insoluble cellular material by *differential centrifugation*. The centrifugal force and duration of centrifugation are adjusted to ensure that the insoluble materials sediment into a pellet. After a starting mixture of a cell homogenate is poured into a tube and spun in a centrifuge, cell organelles such as nuclei collect into a pellet, but the soluble proteins remain in the supernatant. The supernatant fraction still contains a large mixture of proteins, which can be collected by decanting the supernatant and then subjecting it to further purification methods.

RATE-ZONAL CENTRIFUGATION

Based on differences in their mass, proteins can be separated by centrifugation through a solution, usually containing sucrose (an inert sugar), of increasing density called a *density gradient*. When mixtures of proteins are layered on top of a sucrose gradient in a tube and subjected to centrifugation, they migrate down the tube at a rate controlled by the factors that affect the sedimentation constant. The proteins start from a thin zone at the top of the tube and separate into bands, or zones (actually disks), of proteins of different masses. This density-gradient separation technique is called *rate-zonal centrifugation*. Samples are centrifuged just long enough to separate the molecules of interest. If they are centrifuged for too short a time, the molecules will not separate sufficiently. If they are centrifuged much longer than necessary, all the molecules will end up in a pellet at the bottom of the tube.

Although the sedimentation rate is strongly influenced by particle mass, rate-zonal centrifugation is seldom effective in determining *precise* molecular weights because variations in shape also affect sedimentation rate. The exact effects of shape are hard to assess, especially for proteins and single-stranded nucleic acid molecules that can assume many complex shapes. Nevertheless, rate-zonal centrifugation has proved to be the most practical method for separating many different types of polymers and particles. A second density-gradient technique, called *equilibrium density-gradient centrifugation,* is used mainly to separate DNA or organelles.

ELECTROPHORESIS SEPARATES MOLECULES ACCORDING TO THEIR CHARGE:MASS RATIO

Electrophoresis is a technique for separating, or *resolving,* molecules in a mixture under the influence of an applied electric field. Dissolved molecules in an electric field move, or migrate, at a speed determined by their charge:mass ratio. For example, if two molecules have the same mass and shape, the one with the greater net charge will move faster toward an electrode. The separation of small molecules, such as amino acids and nucleotides, is one of the many uses of electrophoresis. In this case, a small drop of sample is deposited on a strip of filter paper or other porous substrate, which is then soaked with a conducting solution. When an electric field is applied at the ends of the strip, small molecules dissolved in the conducting solution move along the strip at a rate corresponding to the magnitude of their charge.

SDS-POLYACRYLAMIDE GEL ELECTROPHORESIS

Because many proteins or nucleic acids that differ in size and shape have nearly identical charge:mass ratios, electrophoresis of these macromolecules in solution results in little or no separation of molecules of different lengths. However, successful separation of proteins and nucleic acids can be accomplished by electrophoresis in various *gels* (semisolid suspensions in water) rather than in a liquid solution. Electrophoretic separation of proteins is most commonly performed in *polyacrylamide gels.* These gels are cast between a pair of glass plates by polymerizing a solution of acrylamide monomers into polyacrylamide chains and simultaneously cross-linking the chains into a semisolid matrix. The *pore size* of a gel can be varied by adjusting the concentrations of polyacrylamide and the cross-linking reagent.

When a mixture of proteins is applied to a gel and an electric current applied, smaller proteins migrate faster than larger proteins through the gel. The rate of movement is influenced by the gel's pore size and the strength of the electric field. The pores in a highly cross-linked polyacrylamide gel are quite small. Such a gel could resolve small proteins and peptides, but large proteins would not be able to move through it.

In what is probably the most powerful technique for resolving protein mixtures, proteins are exposed to the ionic detergent SDS (sodium dodecylsulfate) before and during gel electrophoresis. SDS denatures proteins, causing multimeric proteins to dissociate into their subunits, and all polypeptide chains are forced into extended conformations with similar charge:mass ratios. SDS treatment thus eliminates the effect of differences in shape, so that chain length, which reflects mass, is the sole determinant of the migration rate of proteins in SDS-polyacrylamide electrophoresis. Even chains that differ in molecular weight by less than 10 percent can be separated by this technique. Moreover, the molecular weight of a protein can be estimated by comparing the distance it migrates through a gel with the distances that proteins of known molecular weight migrate.

TWO-DIMENSIONAL GEL ELECTROPHORESIS

Electrophoresis of all cellular proteins through an SDS gel can separate proteins having relatively large differences in molecular weight but cannot resolve proteins having similar molecular weights (e.g., a 41-kDa protein from a 42-kDa protein). To separate proteins of similar mass, another physical characteristic must be exploited. Most commonly, this is electric charge, which is determined by the number of acidic and basic residues in a protein. Two unrelated proteins having similar masses are unlikely to have identical net charges because their sequences, and thus the number of acid and basic residues, are different.

In two-dimensional electrophoresis, proteins are separated in two sequential steps: first by their charge and then by their mass. In the first step, a cell extract is fully denatured by high concentrations (8 M) of urea and then layered on a glass tube filled with polyacrylamide that is saturated with a solution of *ampholytes*, a mixture of polyanionic and polycationic molecules. When placed in an electric field, the ampholytes will separate and form a continuous gradient based on their net charge. The most highly polyanionic ampholytes will collect at one end of the tube, and the most polycationic ampholytes will collect at the other end. This gradient of ampholytes establishes a pH gradient. Charged proteins will migrate through the gradient until they reach their pI, or isoelectric point, the pH at which the net charge of the protein is zero. This technique, called isoelectric focusing (IEF), can resolve proteins that differ by only one charge unit.

Proteins that have been separated on an IEF gel can then be separated in a second dimension based on their molecular weights. To accomplish this, the IEF gel is extruded from the tube and placed lengthwise on a second polyacrylamide gel, this time formed as a slab saturated with SDS. When an electric field is imposed, the proteins will migrate from the IEF gel into the SDS slab gel and then separate according to their mass. The sequential resolution of proteins by their charge and mass can achieve excellent separation of cellular proteins. For example, two-dimensional gels have been very useful in studying the expression of various genes in differentiated cells because as many as 1000 proteins can be resolved simultaneously.

LIQUID CHROMATOGRAPHY

Liquid chromatography, a third commonly used technique to separate mixtures of proteins, nucleic acids, and other molecules, is based on the principle that molecules dissolved in a solution will interact (bind and dissociate) with a solid surface. If the solution is allowed to flow across the surface, then molecules that interact frequently with the surface will spend more time bound to the surface and thus move more slowly than molecules that interact infrequently with the surface. Liquid chromatography is performed in a column packed tightly with spherical beads. The nature of these beads determines whether separation of proteins depends on differences in mass, charge, or binding affinity.

GEL FILTRATION CHROMATOGRAPHY

Proteins that differ in mass can be separated by gel filtration chromatography. In this technique, the column is composed of porous beads made from polyacrylamide, dextran (a bacterial polysaccharide), or agarose (a seaweed derivative). Proteins *flow around* the spherical beads in gel filtration chromatography. However, the surface of the beads is punctured by large holes, and proteins will spend some time within these holes. Because smaller proteins can penetrate into the beads more easily than larger proteins, they travel through a gel filtration column more slowly than larger proteins. (In contrast, proteins migrate *through* the pores in an electrophoretic gel; thus smaller proteins move faster than larger ones.) The total volume of liquid required to elute a protein from the column depends on its mass: the smaller the mass, the greater the elution volume. By use of proteins of known mass, the elution volume can be used to estimate the mass of a protein in a mixture.

ION-EXCHANGE CHROMATOGRAPHY

In a second type of liquid chromatography, called *ion-exchange chromatography*, proteins are separated based on differences in their charge. This technique makes use of specially modified beads whose surfaces are covered by amino groups or carboxyl groups and thus carry either a positive charge (NH_3^+) or a negative charge (COO^-) at neutral pH.

The proteins in a mixture carry various net charges at any given pH. When a solution of a protein mixture flows through a column of positively charged beads, only proteins with a net negative charge (acidic proteins) adhere to the beads; neutral and basic proteins flow unimpeded through the column. The acidic proteins are then eluted selectively by passing a gradient of increasing concentrations of salt through the column. At low salt concentrations, protein molecules and beads are attracted by their opposite charges. At higher salt concentrations, negative salt ions bind to the positively charged beads, displacing the negatively charged proteins. In a gradient of increasing salt concentration, weakly charged proteins are eluted first and highly charged proteins are eluted last. Similarly, a negatively charged column can be used to retain and fractionate positively charged (basic) proteins.

AFFINITY CHROMATOGRAPHY

A third form of chromatography, called *affinity chromatography,* relies on the ability of a protein to bind specifically to another molecule. Columns are packed with beads to which are covalently attached ligand molecules that bind to the protein of interest. Ligands can be enzyme substrates or other small molecules that bind to specific proteins. In a widely used form of this technique, *antibody-affinity chromatography,* the attached ligand is an antibody specific for the desired protein. An affinity column will retain only the proteins that bind the ligand attached to the beads; the remaining proteins, regardless of their charge or mass, will pass through the column without binding to it. The proteins bound to the affinity column then are eluted by adding an excess of ligand or by changing the salt concentration or pH. Obviously, the ability of this technique to separate particular proteins depends on the selection of appropriate ligands.

HIGHLY SPECIFIC ENZYME AND ANTIBODY ASSAYS CAN DETECT INDIVIDUAL PROTEINS

Purification of a protein, or any other molecule, requires a specific assay that can detect the molecule of interest in column fractions or gel bands. An assay capitalizes on some highly distinctive characteristic of a protein: the ability to bind a particular ligand, to catalyze a particular reaction, or to be recognized by a specific antibody. An assay must also be simple and fast in order to minimize errors and the possibility that the protein of interest is denatured or degraded while the assay is performed. The goal of any purification scheme is to isolate sufficient amounts of a given protein for study; thus a useful assay must also be sensitive enough that only a small proportion of the available material is consumed. Many common protein assays require just 10^{-9} to 10^{-12} g of material.

CHROMOGENIC AND LIGHT-EMITTING ENZYME REACTIONS

Many assays are tailored to detect some functional aspect of a protein. For example, enzyme assays are based on the ability to detect the loss of substrate or the formation of product. Many enzyme assays utilize *chromogenic* substrates, which change colour during the course of the reaction. (Some substrates are naturally chromogenic; if they are not, they can be linked to a chromogenic molecule.) Because of the specificity of an enzyme for its substrate, only samples that contain the enzyme will change colour in the presence of a chromogenic substrate and other required reaction components; the rate of the reaction provides a measure of the quantity of enzyme present.

Such chromogenic enzymes also can be fused or chemically linked to an antibody and used to "report" the presence or location of the antigen. Alternatively *luciferase,* an enzyme present in fireflies and some bacteria, can be linked to an antibody. In the presence of ATP and luciferin, this enzyme catalyzes a light-emitting reaction. In either case, after the antibody binds to the protein of interest, substrates of the linked enzyme are added and the appearance of colour or emitted light is monitored.

WESTERN BLOTTING

One of the most powerful methods for detecting a particular protein in a complex mixture combines the superior resolving power of gel electrophoresis, the specificity of antibodies, and the sensitivity of enzyme assays. Called Western blotting, or immunoblotting, this three-step procedure is commonly used to separate proteins and then identify a specific protein of interest. As shown two different antibodies are used in this method, one specific for the desired protein and the other linked to a reporter enzyme.

RADIOISOTOPES ARE INDISPENSABLE TOOLS FOR DETECTING BIOLOGICAL MOLECULES

Since World War II, when radioactive materials first became widely available as byproducts of work in nuclear physics, chemists and biologists have fashioned an almost limitless variety of radioactive chemicals. Today, radioactively labeled precursors of macromolecules greatly simplify many standard biochemical assays and significantly enhance the ability of researchers to follow biochemical events in whole cells as well as in cell extracts. Almost all experimental biology depends on the use of radioactive compounds.

At least one atom in a radiolabeled molecule is present in a radioactive form, called a radioisotope. The presence of a radioisotope does not change the chemical properties of a molecule. For example, enzymes, both in vivo and in vitro, catalyze reactions involving labeled substrates just as readily as those involving nonlabeled substrates. Because radioisotopes emit easily detected particles, the fate of radiolabeled molecules can be traced in cells and cellular extracts.

CHARACTERISTICS OF DIFFERENT RADIOLABELS

The choice of which labeled compound to use in a particular experiment involves several considerations. Some labeled compounds, for instance, are not suitable for studies with whole cells because they do not enter cells. One prominent example is ATP, as well as most other phosphorylated compounds (e.g., glucose 6-phosphate). Although ^{32}P-labeled ATP can contribute phosphorus-32 during RNA and DNA synthesis in a cell-free system, it cannot do so with whole cells, because it never gets into the cells. On the other hand, labeled orthophosphate ($^{32}PO_4^{3-}$) in the medium does enter both bacterial cells and animal cells, and then is incorporated into phosphorylated proteins, nucleotides, and eventually into cellular RNA and DNA.

Hundreds of biological compounds (e.g., amino acids, nucleosides, and numerous metabolic intermediates) labeled with various radioisotopes are commercially available. These preparations vary considerably in their *specific activity*, which is the amount of radioactivity per unit of material, measured in disintegrations per minute (dpm) per millimole. The specific activity of a labeled compound depends on the ratio of unstable potentially radioactive atoms to stable nonradioactive atoms. It also depends on the probability of decay of the radioisotope, indicated by its *half-life*, which is the time required for half the atoms to undergo radioactive decay. In general, the shorter the half-life of a radioisotope, the higher its specific activity.

The specific activity of a labeled compound must be high enough so that sufficient radioactivity is incorporated into cellular molecules to be accurately detected. For example, methionine and cysteine labeled with sulfur-35 (^{35}S) are widely used to label cellular proteins because preparations of these amino acids with high specific activities ($>10^{15}$ dpm/mmol) are available. Likewise, commercial preparations of ^{3}H-labeled nucleic acid precursors have much higher specific activities than the corresponding ^{14}C-labeled preparations. In most experiments, the former are preferable because they allow RNA or DNA to be adequately labeled after a shorter time of incorporation or require a smaller cell sample.

Various phosphate-containing compounds in which every phosphorus atom is the radioisotope phosphorus-32 are readily available. Because of their high specific activity, ^{32}P-labeled nucleotides are routinely used to label nucleic acids in cell-free systems. The radioisotope iodine-125 (^{125}I), which also is available in almost pure form, can be covalently linked to a protein or nucleic acid to yield preparations with a high specific activity. Such attachment of iodine-125 can be achieved enzymatically or chemically and generally does not drastically alter the properties of a macromolecule.

LABELING EXPERIMENTS AND DETECTION OF RADIOLABELED MOLECULES

Depending on the nature of an experiment, labeled compounds are detected by autoradiography, a semiquantitative visual assay, or their radioactivity is measured in an appropriate "counter," a highly quantitative assay that can determine the concentration of a radiolabeled compound in a sample. In some experiments, both types of detection are used.

In autoradiography, a cell or cell constituent is labeled with a radioactive compound and then overlaid with a photographic emulsion sensitive to radiation. Development of the emulsion yields small silver grains whose distribution corresponds to that of the radioactive material. Autoradiographic studies of whole cells have been crucial in determining the intracellular sites where various macromolecules are synthesized and their subsequent movements within cells. For example, when cells are incubated for a short time with [^{3}H]thymidine, a unique DNA precursor, most of the radioactivity is localized to the nucleus, identifying the nucleus as the major site of DNA synthesis. Even after cells are incubated for prolonged periods with [^{3}H]thymidine, virtually all the radioactivity remains in the nucleus, indicating that the DNA remains there. Similarly, the site of synthesis of RNA is revealed by incubating cells for 1 minute with [^{3}H]uridine, a unique RNA precursor; in this case, all the autoradiographic grains are found over the nucleus. After a longer period of incorporation, however, many autoradiographic grains are located over the cytoplasm, indicating that RNA, in contrast to DNA, is transported from the nucleus.

In autoradiographic studies, the ability of the experimenter to localize the site at which the radioisotope is incorporated is affected by the energy of the particles emitted during radioactive disintegrations. For example, the β particles emitted by phosphorus-32 are so energetic that the streaks they make on a photographic emulsion can be as long as 1 mm, much longer than the diameter of individual cells. In contrast, the β particles emitted by tritium create tracks on a photographic emulsion that are only about 0.47 μm long; thus 3H-labeled structures can be located within cells to an accuracy of about 0.5–1.0 mm, or about one-fifth the diameter of the nucleus of mammalian cells. Because tritium emits the least-energetic particles of all the common radioisotopes, it is highly preferred for locating labeled compounds or structures within cells.

Quantitative measurements of the amount of radioactivity in a labeled material are performed with several different instruments. A *Geiger counter* measures ions produced in a gas by the β particles or γ rays emitted from a radioisotope. In a *scintillation counter,* a radiolabeled sample is mixed with a liquid containing a fluorescent compound that emits a flash of light when it absorbs the energy of the β particles or γ rays released during decay of the radioisotope; a phototube in the instrument detects and counts these light flashes. *Phosphorimagers* are used to detect radiolabeled compounds on a surface, storing digital data on the number of decays in dpm per small pixel of surface area. These instruments commonly are used to quantitate radioactive molecules separated by gel electrophoresis and are replacing photographic film for this purpose.

The usual experimental protocol for determining the cellular location of a particular molecule has three steps:

1. A radioactive precursor is incubated with whole cells or cell-free extracts.
2. The cellular constituents then are isolated and purified in various ways.
3. The radioactivity of the various fractions is measured with a counter.

For example, to identify the site of RNA synthesis, cells can be incubated for a short period with [^{3}H]uridine and then subjected to a fractionation procedure to separate the various organelles. The specific activity of the nuclear fraction (dpm/mg protein) is found to be much higher than that of any other organelle fraction, thus confirming the nucleus as the site of RNA synthesis.

A combination of labeling and biochemical techniques and of visual and quantitative detection methods is often employed in labeling experiments. For instance, to identify the major proteins synthesized by a particular cell type, a sample of the cells is incubated with a radioactive amino acid (e.g., ^{35}S-labeled methionine) for a few minutes. The mixture of cellular proteins then is resolved by gel electrophoresis, and the gel is subjected to autoradiography or phosphorimager analysis. The radioactive bands correspond to newly synthesized proteins, which have incorporated the radiolabeled amino acid. Alternatively, the proteins can be resolved by liquid chromatography, and the radioactivity in the eluted fractions determined quantitatively with a counter.

Researchers often use the pulse-chase technique in labeling experiments. In this protocol, a cell sample is exposed to a radiolabeled compound for only a brief period of time (the "pulse"), then washed with buffer to remove the label, and finally incubated with a nonlabeled form of the compound (the "chase"). Pulse-chase experiments are particularly useful for tracing changes in the intracellular location of proteins or the transformation of a metabolite into others over time.

PROTEIN PRIMARY STRUCTURE CAN BE DETERMINED BY CHEMICAL METHODS AND FROM GENE SEQUENCES

The primary structure of a protein is characterized in two ways: by its overall amino acid composition and by its precise amino acid sequence. The amino acid composition of a protein gives the same information as an elemental analysis of a molecule—the types of amino acids present and their abundance but not their linear order. In contrast, the sequence of a protein is like a fingerprint; it uniquely establishes the identity of a protein—the linear order of amino acids. The composition of a protein is easily calculated from the sequence. Two proteins can differ in their sequence but nonetheless have identical amino acid compositions. Composition and sequence are determined by chemical methods based on the ability to cleave the peptide backbone at the peptide bond.

The classic method for determining the amino acid sequence of a protein involves *Edman degradation*. In this procedure the amino group at the N-terminus of a polypeptide is labeled and its amino acid then cleaved from the polypeptide and identified by high-pressure liquid chromatography. The polypeptide is left one residue shorter, with a new amino acid at the N-terminus. The cycle is repeated on the ever shortening polypeptide until all the residues have been identified.

Before about 1985, biologists commonly used the Edman chemical procedure for determining protein sequences. However, recombinant DNA techniques developed in the 1970s and 1980s permit the detection and cloning of the mRNA or gene encoding a specific protein. From the sequence of its mRNA or gene, the protein's amino acid sequence can be deduced. Because sequencing of mRNA and DNA generally is faster than chemical sequencing of proteins, this approach is now the most popular way to determine protein sequences, especially of large proteins. As we describe the complete genome sequences of several organisms have already been determined, and the database of genome sequences from humans and numerous model organisms is expanding rapidly.

TIME-OF-FLIGHT MASS SPECTROMETRY MEASURES THE MASS OF PROTEINS AND PEPTIDES

A powerful technique for measuring the mass of molecules like proteins and peptides is *mass spectrometry*. Mass spectrometry requires a method for ionizing the sample, usually a mixture of peptides or proteins, accelerating the molecular ions, and then detecting the ions. In a laser desorption mass spectrometer, the proteins are mixed with an organic acid and then dried on a metal target. Light from a laser ionizes the proteins, which "fly" down a tube to a detector. Their time of flight is inversely proportional to their mass and directly proportional to the charge on the protein. As little as 1 femtomole of proteins as large as 200,000 MW can be measured with an error of 0.1 percent.

One powerful use of mass spectrometers is to identify a protein from its *peptide mass fingerprint*. A peptide mass fingerprint is a compilation of the molecular weights of peptides that are generated by a specific protease. The molecular weights of the parent protein and its proteolytic fragments are used to search genome databases for any similarly sized protein with identical or similar peptide mass maps. With the increasing availability of genome sequences, this approach has almost eliminated the need to chemically sequence a protein to determine its primary structure.

PEPTIDES WITH A DEFINED SEQUENCE CAN BE SYNTHESIZED CHEMICALLY

Synthetic peptides that are identical with peptides synthesized in vivo are useful experimental tools in studies of proteins and cells. For example, short synthetic peptides of 10–15 residues can function as antigens to trigger production of antibodies in animals. A synthetic peptide can trick the animal into producing antibodies that bind the full-sized, natural protein antigen. As we'll see throughout this book, antibodies are extremely versatile reagents for isolating proteins from mixtures by affinity chromatography, separating and detecting proteins by Western blotting, and localizing proteins in cells by microscopic techniques. Synthetic peptides also have been helpful in elucidating the rules that determine the secondary and tertiary structure of proteins. By systematically varying the sequence of synthetic peptides, researchers have studied the influence of various amino acids on protein conformation.

Peptides are routinely synthesized in a test tube from monomeric amino acids by condensation reactions that form peptide bonds. Peptides are constructed sequentially by coupling the C-terminus of a monomeric amino acid with the N-terminus of the growing peptide. To prevent unwanted reactions involving the amino groups and carboxyl groups of the side chains during the coupling steps, a protecting (blocking) group is attached to the side chains. Without these protecting groups, branched peptides would be generated. In the last steps of synthesis, the side chain–protecting groups are removed and the peptide is cleaved from the resin on which synthesis occurs.

PROTEIN CONFORMATION IS DETERMINED BY SOPHISTICATED PHYSICAL METHODS

In this chapter we have emphasized that protein function

is derived from protein structure. Thus, to figure out how a protein works, its three-dimensional structure must be known. Determining a protein's conformation requires sophisticated physical methods and complex analyses of the experimental data. We briefly describe three methods used to generate three-dimensional models of proteins.

X-RAY CRYSTALLOGRAPHY

The use of x-ray crystallography to determine the three-dimensional structures of proteins was pioneered by Max Perutz and John Kendrew in the 1950s. To date, the detailed three-dimensional structures of more than 8000 proteins have been established by this technique in which beams of x-rays are passed through a crystal of protein. The wavelengths of x-rays are about 0.1–0.2 nanometer (nm), short enough to resolve the atoms in the protein crystal. Atoms in the protein crystal scatter the x-rays, which produce a diffraction pattern of discrete spots when they are intercepted by photographic film. Such patterns are extremely complex; as many as 25,000 diffraction spots can be obtained from a small protein. Elaborate calculations and modifications of the protein (such as binding of heavy metals) must be made to interpret the diffraction pattern and to solve the structure of the protein. The process is analogous to reconstructing the precise shape of a rock from the ripples it creates in a pond.

CRYOELECTRON MICROSCOPY

Although some proteins readily crystallize, obtaining crystals of others—particularly large multisubunit proteins—requires a time-consuming trial-and-error effort to find just the right conditions. Low-resolution views of such difficult-to-crystallize proteins can be obtained by electron microscopy. In this technique, a protein sample is rapidly frozen in liquid helium to preserve its structure and then examined in the frozen, hydrated state in a *cryoelectron microscope.* Pictures are recorded on film using a low dose of electrons to prevent radiation-induced damage to the structure. Sophisticated computer programmes analyse the images and reconstruct the protein's structure in three dimensions. With recent advances in cryoelectron microscopy, this technique can generate molecular models that compare with those generated by x-ray crystallography.

NMR SPECTROSCOPY

The three-dimensional structures of small proteins containing up to about 200 amino acids can be studied with nuclear magnetic resonance (NMR) spectroscopy. In this technique, a concentrated protein solution is placed in a magnetic field and the effects of different radio frequencies on the resonances of different atoms are measured. However, the behavior of any atom is influenced by neighboring atoms in adjacent residues; the closely spaced residues are more perturbed than distant residues. From the magnitude of the effect, the distances between residues can be calculated; these distances then are used to generate a model of the three-dimensional structure of the protein. Although NMR does not require crystallization of a protein, a definite advantage, this technique is limited to proteins smaller than about 20 kDa. However, NMR analysis also can be applied to protein domains, which tend to be small enough for this technique and often can be obtained as stable structures.

Chapter 4

Nucleic Acids: the Genetic Code and the Synthesis of Macromolecules

The extraordinary versatility of proteins in catalyzing chemical reactions and building cellular structures was featured in the previous chapter. In this chapter we consider the nucleic acids, the molecules tha

1. contain the information prescribing amino acid sequence in proteins
2. serve in the several cellular structures that choose, and then link into the correct order, the amino acids of a protein chain. Deoxyribonucleic acid (DNA) is the storehouse, or cellular library, that contains all the information required to build the cells and tissues of an organism. The exact duplication of this information in any species from generation to generation assures the genetic continuity of that species. The information is arranged in units identified by classical geneticists from Gregor Mendel through Thomas Hunt Morgan, and known now as genes, hereditary units controlling identifiable traits of an organism. In the process of transcription, the information stored in DNA is copied into ribonucleic acid (RNA), which has three distinct roles in protein synthesis. Messenger RNA (mRNA) carries the

instructions from DNA that specify the correct order of amino acids during protein synthesis. The remarkably accurate, stepwise assembly of amino acids into proteins occurs by translation of mRNA. In this process, the information in mRNA is interpreted by a second type of RNA called transfer RNA (tRNA) with the aid of a third type of RNA, ribosomal RNA (rRNA), and its associated proteins. As the correct amino acids are brought into sequence by tRNAs, they are linked by peptide bonds to make proteins.

Discovery of the structure of DNA in 1953 and the subsequent elucidation of the steps in the synthesis of DNA, RNA, and protein are the monumental achievements of the early days of molecular biology. To understand how DNA directs synthesis of RNA, which then directs assembly of proteins — the so-called *central dogma* of molecular biology — we first discuss the building blocks composing DNA and RNA, the cell's two primary nucleic acids, and their general structures. We then introduce the basic mechanisms of DNA and RNA chain synthesis, including a brief discussion of gene structure, which reveals why molecular processing is required to make functional RNA molecules. Next we outline the roles of mRNA, tRNA, and rRNA in protein synthesis. The chapter closes with a detailed description of the components and biochemical steps in the formation of proteins. Since the events of macromolecular synthesis are so central to all biological functions — growth control, differentiation, and the specialized chemical and physical properties of cells — they will arise again and again in later chapters. A firm grasp of the fundamentals of DNA, RNA, and protein synthesis is necessary to follow the subsequent discussions without difficulty.

STRUCTURE OF NUCLEIC ACIDS

DNA and RNA have great chemical similarities. In their *primary structures* both are linear polymers (multiple chemical units) composed of monomers (single chemical units), called nucleotides. Cellular RNAs range in length from less than one hundred to many thousands of nucleotides. Cellular DNA molecules can be as long as several hundred million nucleotides. These large DNA units in association with proteins can be stained with dyes and visualized in the light microscope as chromosomes.

POLYMERIZATION OF NUCLEOTIDES FORMS NUCLEIC ACIDS

DNA and RNA each consists of only four different nucleotides. All nucleotides have a common structure: a *phosphate* group linked by a phosphoester bond to a *pentose* (a five-carbon sugar molecule) that in turn is linked to an organic *base*. In RNA, the pentose is *ribose;* in DNA, it is *deoxyribose*. The only other difference in the nucleotides of DNA and RNA is that one of the four organic bases differs between the two polymers. The bases adenine, guanine, and cytosine are found in both DNA and RNA; thymine is found only in DNA, and uracil is found only in RNA. The bases are often abbreviated A, G, C, T, and U, respectively. For convenience the single letters are also used when long sequences of nucleotides are written out.

The base components of nucleic acids are heterocyclic compounds with the rings containing nitrogen and carbon. Adenine and guanine are purines, which contain a pair of fused rings; cytosine, thymine, and uracil are pyrimidines, which contain a single ring. The acidic character of nucleotides is due to the presence of phosphate, which dissociates at the pH found inside cells, freeing hydrogen ions and leaving the phosphate negatively charged. Because these charges attract proteins, most nucleic acids in cells are associated with proteins. In nucleotides, the 12 carbon atom of the sugar (ribose or deoxyribose) is attached to the nitrogen at position 9 of a purine (N_9) or at position 1 of a pyrimidine (N_1).

Cells and extracellular fluids in organisms contain small concentrations of nucleosides, combinations of a base and a sugar without a phosphate. Nucleotides are nucleosides that have one, two, or three phosphate groups esterified at the 52 hydroxyl. *Nucleoside monophosphates* have a single esterified phosphate *diphosphates* contain a prophosphate group

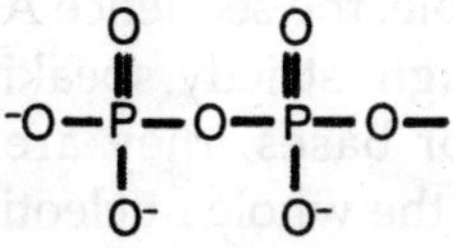

and *triphosphates* have a third phosphate. As we will see later, the nucleoside triphosphates are used in the synthesis of nucleic acids. However, these compounds also serve many other functions in the cell: ATP, for example, is the most widely used energy carrier in the cell, and GTP plays crucial roles in intracellular signaling and acts as an energy reservoir, particularly in protein synthesis.

When nucleotides polymerize to form nucleic acids, the hydroxyl group attached to the 32 carbon of a sugar of one nucleotide forms an ester bond to the phosphate of another nucleotide, eliminating a molecule of water:

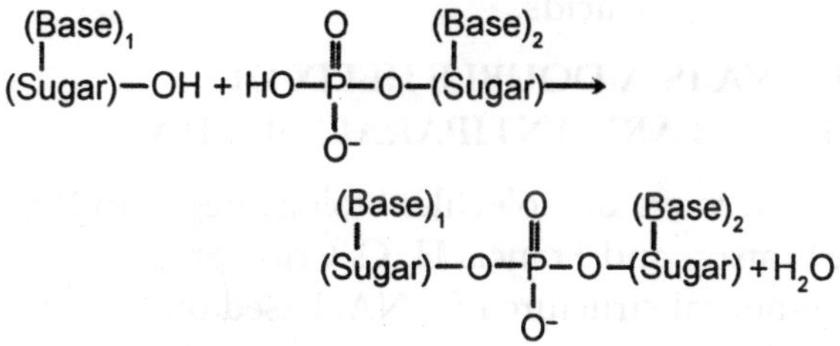

This condensation reaction is similar to that in which a peptide bond is formed between two amino acids. Thus a single nucleic acid strand is a phosphate-pentose polymer (a polyester) with purine and pyrimidine bases as side groups. The links between the nucleotides are called phosphodiester bonds. Like a polypeptide, a nucleic acid strand has an end-to-end chemical orientation: the 52 *end* has a free hydroxyl or phosphate group on the 52 carbon of its terminal sugar; the 32 *end* has a free hydroxyl group on the 32 carbon of its terminal sugar. This directionality, plus the fact that synthesis proceeds 52 to 32, has given rise to the convention that polynucleotide sequences are written and read in the 52 $\rightarrow$ 32 direction (from left to right); for example, the sequence AUG is assumed to be (52) AUG(32). (Although, strictly speaking, the letters A, G, C, T, and U stand for bases, they are also often used in diagrams to represent the whole nucleotides containing these bases.) The 52 $\rightarrow$ 32 directionality of a nucleic acid strand is an extremely important property of the molecule.

The linear sequence of nucleotides linked by phosphodiester bonds constitutes the primary structure of nucleic acids. As we discuss in the next section, polynucleotides can twist and fold into three-dimensional conformations stabilized by noncovalent bonds; in this respect, they are similar to polypeptides. Although the primary structures of DNA and RNA are generally similar, their conformations are quite different. Unlike RNA, which commonly exists as a single polynucleotide chain, or strand, DNA contains two intertwined polynucleotide strands. This structural difference is critical to the different functions of the two types of nucleic acids.

NATIVE DNA IS A DOUBLE HELIX OF COMPLEMENTARY ANTIPARALLEL CHAINS

The modern era of molecular biology began in 1953 when James D. Watson and Francis H. C. Crick proposed correctly the double-helical structure of DNA, based on the analysis of x-ray diffraction patterns coupled with careful model building. A closer look at the "thread of life," as the DNA molecule is sometimes called, shows why the discovery of its basic structure suggests its function.

DNA consists of two associated polynucleotide strands that wind *together* through space to form a structure often described as a double helix. The two sugar-phosphate backbones are on the outside of the double helix, and the bases project into the interior. The adjoining bases in each strand stack on top of one another in parallel planes. The orientation of the two strands is antiparallel; that is, their 52 → 32 directions are opposite. The strands are held in precise register by a regular base-pairing between the two strands: A is paired with T through two hydrogen bonds; G is paired with C through three hydrogen bonds. This *base-pair complementarity* is a consequence of the size, shape, and chemical composition of the bases. The presence of thousands of such hydrogen bonds in a DNA molecule contributes greatly to the stability of the double helix. Hydrophobic and van der Waals interactions between the stacked adjacent base pairs also contribute to the stability of the DNA structure.

To maintain the geometry of the double-helical structure a larger purine (A or G) must pair with a smaller pyrimidine (C or T). In natural DNA, A almost always hydrogen bonds with T and G with C, forming A·T and G·C base pairs often called *Watson-Crick base pairs*. Two polynucleotide strands, or regions thereof, in which all the nucleotides form such base pairs are said to be complementary. However, in theory and in synthetic DNAs other interactions can occur. For example, a guanine (a purine) could theoretically form hydrogen bonds with a thymine (a pyrimidine), causing only a minor distortion in the helix. The space available in the helix also would allow pairing between the two pyrimidines cytosine and thymine. Although the nonstandard G·T and C·T base pairs are normally not found in DNA, G·U base pairs are quite common in double-helical regions that form within otherwise single-stranded RNA.

Two polynucleotide strands can, in principle, form either a right-handed or a left-handed helix. Because the geometry of the sugar-phosphate backbone is more compatible with the former, natural DNA is a right-handed helix. The x-ray diffraction pattern of DNA indicates that the stacked bases are regularly spaced 0.34 nm apart along the helix axis. The helix makes a complete turn every 3.4 nm; thus there are about 10 pairs per turn. This is referred to as the *B form* of DNA, the normal form present in most DNA stretches in cells. On the outside of B-form DNA, the spaces between the intertwined strands form two helical grooves of different widths described as the *major* groove and the *minor* groove. Consequently, part of each base is accessible from outside the helix to both small and large molecules that bind to the DNA by contacting chemical groups within the grooves. These two binding surfaces of the DNA molecule are used by different classes of DNA-binding proteins.

In addition to the major B form of DNA, three additional structures have been described. In very low humidity, the crystallographic structure of B DNA changes to the *A form;* RNA-DNA and RNA-RNA helices also exist in this form. The A form is more compact than the B form, having 11 bases per turn, and the stacked bases are tilted. Short DNA molecules composed of alternating purine-pyrimidine nucleotides (especially Gs and Cs) adopt an alternative left-handed configuration instead of the normal right-handed helix. This structure is called *Z DNA* because the bases seem to zigzag when viewed from the side. It is entirely possible that both A-form and Z-form stretches of DNA exist in cells.

Finally, a triple-stranded DNA structure can also exist at least in the test tube, and possibly during recombination and DNA repair. For example, when synthetic polymers of poly(A) and polydeoxy(U) are mixed, a three-stranded structure is formed. Further, long homopolymeric stretches of DNA composed of C and T residues in one strand and A and G residues in the other can be targeted by short matching lengths of poly(C+T). The synthetic oligonucleotide can insert as a third strand, binding in a sequence-specific manner by so-called *Hoogsteen base pairs*. Specific cleavage of the DNA at the site where the triple helix ends can be achieved by attaching a chemical cleaving agent (e.g., Fe^{2+}-EDTA) to the short oligodeoxynucleotide that makes up the third strand. Such reactions may be useful in studying site-specific DNA damage in cells.

By far the most important modifications in standard B-form DNA come about as a result of protein binding to specific

DNA sequences. Although the multitude of hydrogen and hydrophobic bonds between the polynucleotide strands provide stability to DNA, the double helix is somewhat flexible about its long axis. Unlike the α helix in proteins, there are no hydrogen bonds between successive residues in a DNA strand. This prop- erty allows DNA to bend when complexed with a DNA-binding protein. Crystallographic analyses of proteins bound to particular regions of DNA have conclusively demonstrated departures from the standard B-DNA structure in protein-DNA complexes. The specific DNA-protein contacts that occur in these tightly bound complexes have the ability both to untwist the DNA and to bend the axis of the helix. Although DNA in cells likely exists in the B form most of the time, particular regions bound to protein clearly depart from the standard conformation.

DNA CAN UNDERGO REVERSIBLE STRAND SEPARATION

In DNA replication and in the copying of RNA from DNA, the strands of the helix must separate at least temporarily. As we discuss later, during DNA synthesis two new strands are made (one copied from each of the original strands), resulting in two double helices identical with the original one. In the case of copying the DNA template to make RNA, the RNA is released and the two DNA strands reassociate with each other.

The unwinding and separation of DNA strands, referred to as denaturation, or "melting," can be induced experimentally. For example, if a solution of DNA is heated, the thermal energy increases molecular motion, eventually breaking the hydrogen bonds and other forces that stabilize the double helix, and the strands separate. This melting of DNA changes its absorption of ultraviolet (UV) light (in the 260-nm range), which is routinely used to measure DNA concentration because of the high absorbance of UV light by nucleic acid bases. Native double-stranded DNA absorbs about one-half as much light at 260 nm as does the equivalent amount of single-stranded DNA. Thus, as DNA denatures, its absorption of UV light increases. Near the denaturation temperature, a small increase in temperature causes an abrupt, near simultaneous, loss of the multiple, weak, cooperative interactions holding the two strands together, so that denaturation rapidly occurs throughout the entire length of the DNA.

The *melting temperature*, T_m, at which the strands of DNA will separate depends on several factors. Molecules that contain a greater proportion of G·C pairs require higher temperatures to denature because the three hydrogen bonds in G·C pairs make them more stable than A·T pairs with two hydrogen bonds. Indeed, the percentage of G·C base pairs in a DNA sample can be estimated from its T_m. In addition to heat, solutions of low ion concentration destabilize the double helix, causing it to melt at lower temperatures. DNA is also denatured by exposure to other agents that destabilize hydrogen bonds, such as alkaline solutions and concentrated solutions of formamide or urea:

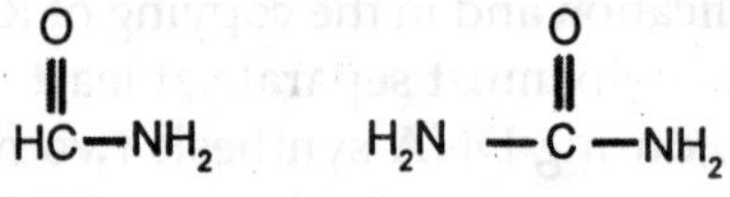

Formamide Urea

The single-stranded DNA molecules that result from denaturation form random coils without a regular structure. Lowering the temperature or increasing the ion concentration causes the two complementary strands to reassociate into a perfect double helix. The extent of such *renaturation* is dependent on time, the DNA concentration, and the ionic content of the solution. Two DNA strands not related in sequence will remain as random coils and will not renature and, most important, will not greatly inhibit complementary DNA partner strands from finding each other. Denaturation and renaturation of DNA are the basis of nucleic acid hybridization, a powerful technique used to study the relatedness of two DNA samples and to detect and isolate specific DNA molecules in a mixture containing numerous different DNA sequences.

MANY DNA MOLECULES ARE CIRCULAR

All prokaryotic genomic DNAs and many viral DNAs are circular molecules. Circular DNA molecules also occur in mitochondria, which are present in almost all eukaryotic cells, and in chloroplasts, which are present in plants and some unicellular eukaryotes.

Each of the two strands in a circular DNA molecule forms a closed structure without free ends. Just as is the case for linear DNA, elevated temperatures or alkaline pH destroy the hydrogen bonds and other interactions that stabilize double-helical circular DNA molecules. Unlike linear DNA, however, the two strands of circular DNA cannot unwind and separate; attempts to melt such DNA result in an interlocked, tangled mass of single-stranded DNA.

Only if a native circular DNA is *nicked* (i.e., one of the strands is cut), will the two strands unwind and separate when the molecule is denatured. In this case, one of the separated strands is circular, and the other is linear. Nicking of circular DNA occurs naturally during DNA replication and can be induced experimentally with a low concentration of deoxyribonuclease (a DNA-degrading enzyme), so that only a single phosphodiester bond in the molecule is cleaved. The study of circular DNA molecules lacking free ends first uncovered the complicated geometric shape changes that the double-stranded DNA molecule must undergo when the strands are not free to separate.

LOCAL UNWINDING OF DNA INDUCES SUPERCOILING

So far we have described DNA as a long regular helical structure that can have local perturbations, especially due to protein binding. In addition, when the two ends of a DNA molecule are fixed, the molecule exhibits a superstructure under certain conditions. This occurs when the base pairing is interrupted and a local region unwinds. The stress induced by unwinding is relieved by twisting of the double helix on itself, forming supercoils. Unwinding and subsequent supercoiling occurs during replication, transcription, and binding of many proteins to circular DNAs or to long DNA loops whose ends are fixed within eukaryotic chromosomes. Supercoiling is recognized and regulated by enzymes called topoisomerases. As discussed in later chapters, these enzymes have an important role in both DNA replication and the transcription of DNA into RNA.

RNA MOLECULES EXHIBIT VARIED CONFORMATIONS AND FUNCTIONS

As noted earlier, the primary structure of RNA is generally similar to that of DNA; however, the sugar component (ribose) of RNA has an additional hydroxyl group at the 22 position and thymine in DNA is replaced by uracil in RNA. The hydroxyl group on C_2 of ribose makes RNA more chemically labile than DNA and provides a chemically·reactive group that takes part in RNA-mediated enzymatic events. As a result of this lability, RNA is cleaved into mononucleotides by alkaline solution, whereas DNA is not. Like DNA, RNA is a long polynucleotide that can be double-stranded or single-stranded, linear or circular. It can also participate in a hybrid helix composed of one RNA strand and one DNA strand; this hybrid has a slightly different conformation than the common B form of DNA.

Unlike DNA, which exists primarily in a single, very long three-dimensional structure, the double helix, the various types of RNA exhibit different conformations. Differences in the sizes and conformations of the various types of RNA permit them to carry out specific functions in a cell. The simplest secondary structures in single-stranded RNAs are formed by pairing of complementary bases. "Hairpins" are formed by pairing of bases within ≈5 – 10 nucleotides of each other, and "stem-loops" by pairing of bases that are separated by ≈50 to several hundred nucleotides. These simple folds can cooperate to form more complicated tertiary structures, one of which is termed a "pseudoknot".

As discussed in detail later, tRNA molecules adopt a well-defined three-dimensional architecture in solution that is crucial in protein synthesis. Larger rRNA molecules also have locally well defined three-dimensional structures, with more flexible links in between. Secondary and tertiary structures also have been recognized in mRNA, particularly near the ends of molecules. These recently discovered structures are under active study. Clearly, then, RNA molecules are like proteins in that they have structured domains connected by less structured, flexible stretches.

The folded domains of RNA molecules not only are structurally analogous to the α helices and β strands found in proteins, but in some cases also have catalytic capacities. Such catalytic RNAs, called ribozymes, can cut RNA chains. Some RNA domains also can catalyze *RNA splicing,* a remarkable process in which an internal RNA sequence, an intron, is cut and removed and the two resulting chains, the exons, are sealed together. This process occurs during formation of the majority of functional mRNA molecules in eukaryotic cells, and also occurs in bacteria and archaea. Remarkably, some RNAs carry out *self-splicing,* with the catalytic activity residing in the intron sequence. As noted later in this chapter, rRNA is thought to play a catalytic role in the formation of peptide bonds during protein synthesis.

In this chapter, we focus on the functions of mRNA, tRNA, and rRNA in gene expression — the process of getting the information in DNA converted into proteins. In later chapters we will encounter other RNAs, often associated with proteins, that participate in other cell functions.

SYNTHESIS OF BIOPOLYMERS: RULES OF MACROMOLECULAR CARPENTRY

According to the central dogma of molecular biology, DNA directs the synthesis of RNA, and RNA then directs the synthesis of proteins. However, the one-way flow of information posited by the central dogma — that is, DNA → RNA → protein — does not reflect the role of proteins in facilitating the information flow. A more accurate way of representing the relationship between the synthesis of DNA, RNA, and proteins in all cells would look like

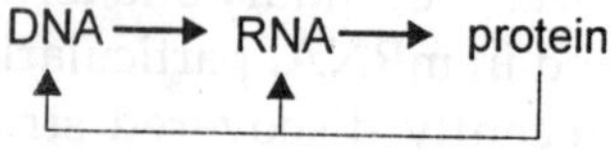

indicating that special proteins catalyze the synthesis of both RNA and DNA.

Interestingly, although the primary and higher-order structures of nucleic acids and proteins differ radically, several common principles apply to the synthesis of both types of macromolecules. Research over the past 30 years has revealed that giant molecular machines, composed largely of proteins, assemble nucleotides into RNA and DNA and amino acids into proteins. Later in this chapter and in subsequent chapters, we examine in detail the molecular machines that synthesize nucleic acids and proteins. To set the stage for these more detailed discussions, we review four general rules that govern the synthesis of both polypeptide and polynucleotide chains:

- *Proteins and nucleic acids are made up of a limited number of different monomeric building blocks.* Although the number of amino acids is theoretically limitless, and several dozen have been identified as metabolic products in various organisms, only 20 different amino acids are used in making proteins. Likewise, only five nitrogenous bases are used to construct RNA and DNA in cells. Cell-free enzyme preparations and cells in culture can be "fooled" into incorporating chemical relatives of the five bases or 20 amino acids, but this almost never happens in nature.

- *The monomers are added one at a time.* Biological polymers theoretically could be built by aligning their component monomeric units in the correct order on a template, or mold, and *simultaneously* fusing them all. But cells do not use this mechanism. Rather, the assembly of proteins and nucleic acids occurs through the step-by- step addition of monomers.

- *Each polypeptide and polynucleotide chain has a specific starting point, and growth proceeds in one direction to a fixed terminus.* Synthesis of chains begins and ends at well-defined "start" and "stop" signals. Monomer addition proceeds from the amino (NH_2-) terminus to the carboxyl (COOH-) terminus in proteins, and from the 52 end to the 32 end in nucleic acids.
- *The primary synthetic product is often modified.* The initially synthesized form of a polypeptide or polynucleotide chain is often inactive or incomplete and undergoes modification to yield an active molecule. Protein chains are often shortened, and RNA chains almost always are; DNA chains are made in pieces and then linked together. Distant segments of long RNA chains may be spliced together, eliminating the intervening sequence (intron). A similar process occurs in some proteins. Polypeptide chains sometimes are cross-linked by covalent bonds to form a functional protein. Finally, certain chemical groups may be added either during the formation of the chain or after its synthesis is complete: e.g., methyl groups to specific sites in DNA, RNA, and proteins; phosphate groups and a wide variety of oligosaccharides to proteins. The details of this macromolecular carpentry are discussed in later chapters.

NUCLEIC ACID SYNTHESIS

The ordered assembly of deoxyribonucleotides into DNA and of ribonucleotides into RNA involves somewhat simpler cellular mechanisms than the correct assembly of the amino acids in a protein chain. Here we consider a few general principles governing the formation of polynucleotide chains in cells and briefly discuss some properties of the enzymes that carry out such synthesis. We also describe the steps in the production of mRNA and examine how and why this process differs in bacteria and eukaryotes. Later chapters cover the mechanism of DNA replication and its control during cell growth and division, and the mechanism and the control of the synthesis of specific mRNAs during differentiation.

BOTH DNA AND RNA CHAINS ARE PRODUCED BY COPYING OF TEMPLATE DNA STRANDS

The regular pairing of bases in the double-helical DNA structure suggested to Watson and Crick a mechanism of DNA synthesis. Their proposal that new strands of DNA are synthesized by copying of parental strands of DNA has proved to be correct.

The DNA strand that is copied to form a new strand is called a template. The information in the template is preserved: although the first copy has a complementary sequence, not an identical one, a copy of the copy produces the original (template) sequence again. In the replication of a double-stranded, or *duplex,* DNA molecule, both original (parental) DNA strands are copied. When copying is finished, the two new duplexes, each consisting of one of the two original strands plus its copy, separate from each other. In some viruses, single-stranded RNA molecules function as templates for synthesis of complementary RNA or DNA chains. However, the vast majority of RNA and DNA in cells is synthesized from preexisting duplex DNA.

NUCLEIC ACID STRANDS GROW IN THE 52 → 32 DIRECTION

All RNA and DNA synthesis, both cellular and viral, proceeds in the same chemical direction: from the 52 (phosphate) end to the 32 (hydroxyl) end. Nucleic acid chains are assembled from 52 triphosphates of ribonucleosides or deoxyribonucleosides. Strand growth is energetically unfavorable but is driven by the energy available in the triphosphates. The α phosphate of the incoming nucleotide attaches to the 32 hydroxyl of the ribose (or deoxyribose) of the preceding residue to form a phosphodiester bond, releasing a pyrophosphate (PP_i). The equilibrium of the reaction is driven further toward chain elongation by pyrophosphatase, which catalyzes the cleavage of PP_i into two molecules of inorganic phosphate.

RNA POLYMERASES CAN INITIATE STRAND GROWTH BUT DNA POLYMERASES CANNOT

The enzymes that copy (replicate) DNA to make more DNA are DNA polymerases; those that copy (transcribe) DNA to form RNA are RNA polymerases. Because the two DNA strands are complementary, rather than identical, transcription of a particular DNA segment theoretically could yield two mRNAs with different sequences and hence different protein-coding potentials. Generally, only one strand of the duplex in a particular DNA segment gives rise to usable information when transcribed into mRNA. In unusual cases, though, limited sections of DNA encode proteins on both strands.

An RNA polymerase can find an appropriate initiation site on duplex DNA; bind the DNA; temporarily "melt," or separate, the two strands in that region; and begin generating a new RNA strand. As discussed the location and regulated use of transcription start sites to produce mRNA requires many dozens of proteins in eukaryotes and several proteins even in bacteria. The nucleotide at the 5 2 terminus of a growing RNA strand is chemically distinct from the nucleotides within the strand in that it retains all three phosphate groups. When an additional nucleotide is added to the 32 end of the growing strand, only the α-phosphate is retained; the β and γ phosphates are lost as pyrophosphate, which is subsequently hydrolyzed to yield 2 molecules of inorganic phosphate.

Unlike RNA polymerases, DNA polymerases cannot initiate chain synthesis de novo; instead, they require a short, preexisting RNA or DNA strand, called a primer, to begin chain growth. With a primer base-paired to the template strand, a DNA polymerase adds nucleotides to the free hydroxyl group at the 32 end of the primer:

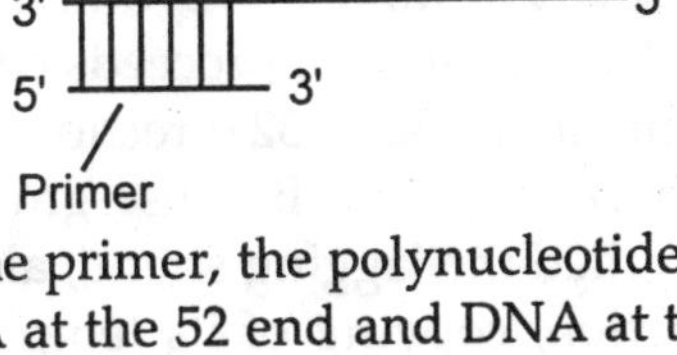

If RNA is the primer, the polynucleotide copied from the template is RNA at the 52 end and DNA at the 32 end.

Both prokaryotic and eukaryotic cells have several different types of DNA polymerases. Some polymerases participate in making new DNA to prepare for cell division; other polymerases serve in the repair and recombination of DNA molecules.

REPLICATION OF DUPLEX DNA REQUIRES ASSEMBLY OF MANY PROTEINS AT A GROWING FORK

Because duplex DNA consists of two intertwined strands, the base-pair copying of each strand requires unwinding of the original duplex, which is accomplished by specific "unwinding proteins" called helicases. As noted earlier, local unwinding of duplex DNA produces torsional stress, leading to formation of supercoils, which are removed by topoisomerases. The action of all these proteins produces a moving, highly specialized region of the DNA called the growing fork, at which DNA polymerase carries out nucleotide addition. In order for DNA polymerase to move along and copy a duplex DNA, helicase must sequentially unwind the duplex and topoisomerase must remove the supercoils that form.

DNA replication begins with creation of a growing fork by a protein or proteins that have helicase activity and unwind a short section of parental DNA. A specialized RNA polymerase then forms short RNA primers complementary to the unwound template strands. Each such primer, still bound to its complementary DNA strand, is then elongated by DNA polymerase, thereby forming a new daughter strand. One final major complication in the operation of a DNA growing fork is that although the two strands of the parental duplex are antiparallel, nucleo-tides can be added to the growing new strands only in the 52 → 32 direction. Synthesis of one daughter strand, called the leading strand, proceeds continuously from a single RNA primer in the 52 → 32 direction, *the same direction as movement of the growing fork*. Because growth of the other daughter strand, called the lagging strand, also must occur in the 52 → 32 direction, copying of its template strand must somehow occur in the *opposite* direction from the movement of the growing fork. A cell accomplishes this feat by producing additional short RNA primers every 1000 bases or so on the second parental strand, as more of the strand is exposed by unwinding. Each of these primers, base-paired to their template strand, is elongated in the 52 $\rightarrow$ 32 direction, forming discontinuous segments called Okazaki fragments after their discoverer Reiji Okazaki. The RNA primer of each Okazaki fragment is removed and replaced by DNA chain growth from the neighboring Okazaki fragment; finally an enzyme called *DNA ligase* joins the adjacent fragments. At least 30 proteins participate in the formation and operation of a DNA growing fork.

ORGANIZATION OF GENES IN DNA DIFFERS IN PROKARYOTES AND EUKARYOTES

Having outlined the principles governing the stepwise assembly of polynucleotides, we now focus briefly on the large-scale arrangement of information in DNA and how this arrangement dictates the requirements for RNA manufacture so that information transfer goes smoothly. The simplest definition of a gene is a "unit of DNA that contains the information to specify synthesis of a single polypeptide chain." The number of genes in cells varies widely, with the simpler non-nucleated prokaryotic cells having far fewer genes than eukaryotic cells. The vast majority of genes carry information to build protein molecules, and it is the RNA copies of such *protein-coding genes* that are the mRNA molecules of cells. In recent years, the entire sequence of the DNA genome of several organisms has been determined, providing direct evidence for large differences in their protein-coding capacity.

The most common arrangement of protein-coding genes in all prokaryotes has a powerful and appealing logic: genes devoted to a single metabolic goal, say, the synthesis of the amino acid tryptophan, are most often found in a contiguous array in the DNA. This gene order makes it possible to produce a continuous strand of mRNA that carries the message for a related series of enzymes devoted to making tryptophan. Each section of the mRNA represents the unit (or gene) that instructs the protein-synthesizing apparatus to make a particular protein. Such an arrangement of genes in a functional group is called an operon, because it operates as a unit from a single transcription start site. In prokaryotic DNA the genes are closely packed with very few noncoding gaps, and the DNA is transcribed directly into colinear mRNA, which then is translated into protein, even while stretches of the mRNA closer to the 32 end are still being produced.

This economic clustering of genes devoted to a single metabolic function does not occur in eukaryotes, even simple ones like yeasts that can be metabolically similar to bacteria. Rather, eukaryotic genes, even those devoted to a single pathway, are most often physically separated in the DNA, sometimes even being located on different chromosomes. Each gene is transcribed from its own start site, producing one mRNA, which generally is translated to yield a single protein. Moreover, when researchers first compared the nucleotide sequences of eukaryotic mRNAs with the DNAs encoding them, they were astounded to find that the uninterrupted protein-coding sequence of a given mRNA was broken up (discontinuous) in its corresponding section of DNA. They concluded that the eukaryotic gene existed in pieces of coding sequence, the *exons,* separated by non-protein-coding segments, the *introns.* This astonishing finding, first discovered in viruses that infect eukaryotic cells, implied that the long initial RNA copy, called the primary transcript, the entire copied DNA sequence, had to be clipped apart to remove the introns and then carefully stitched back together to produce many mRNAs of eukaryotic cells.

EUKARYOTIC PRIMARY RNA TRANSCRIPTS ARE PROCESSED TO FORM FUNCTIONAL MRNAS

In prokaryotic cells, which have no nuclei, translation of an mRNA into protein can begin from the 52 end of the mRNA even while the 32 end is still being copied from DNA. Thus, transcription and translation can occur concurrently. In eukaryotic cells, however, not only is the nucleus separated from the cytoplasm where protein synthesis occurs, but the primary RNA transcript of a protein-coding gene must undergo several modifications, collectively termed RNA processing, that yield a functional mRNA. This mRNA then must be transported to the cytoplasm before it can be translated into protein. Thus, transcription and translation cannot occur concurrently in eukaryotic cells.

The initial steps in processing of all eukaryotic primary RNA transcripts occur at the two ends, and these modifications are retained in mRNAs. To the initiating (52) nucleotide of the primary transcript is added the *52 cap,* which may serve to protect mRNA from enzymatic degradation. This modification occurs before transcription is complete, so the 52 cap is present in the primary transcript. Processing at the 32 end of the primary transcript involves cleavage by an endonuclease to yield a free 32 -hydroxyl group to which a string of adenylic acid residues is added by an enzyme called *poly(A) polymerase.* The resulting poly(A) tail contains 100 – 250 bases, being shorter in yeasts and invertebrates than in vertebrates. Poly(A) polymerase is part of a complex of proteins that adds the poly(A) tail. This complex does not require a template and can determine the correct number of A residues to add in each species.

The final step in the processing of many different eukaryotic mRNA molecules is splicing: the internal cleavage of the RNA transcript to excise the introns, followed by ligation of the coding exons. Many eukaryotic mRNAs also contain noncoding regions at each end; these are referred to as the 52 and 32 *untranslated regions* (UTRs). We examine the cellular machinery for carrying out processing of mRNA, as well as tRNA and rRNA.

THE THREE ROLES OF RNA IN PROTEIN SYNTHESIS

Although DNA stores the information for protein synthesis and RNA carries out the instructions encoded in DNA, most biological activities are carried out by proteins.

The accurate synthesis of proteins thus is critical to the proper functioning of cells and organisms. We saw that the linear order of amino acids in each protein determines its three-dimensional structure and activity. For this reason, assembly of amino acids in their correct order, as encoded in DNA, is the key to production of functional proteins.

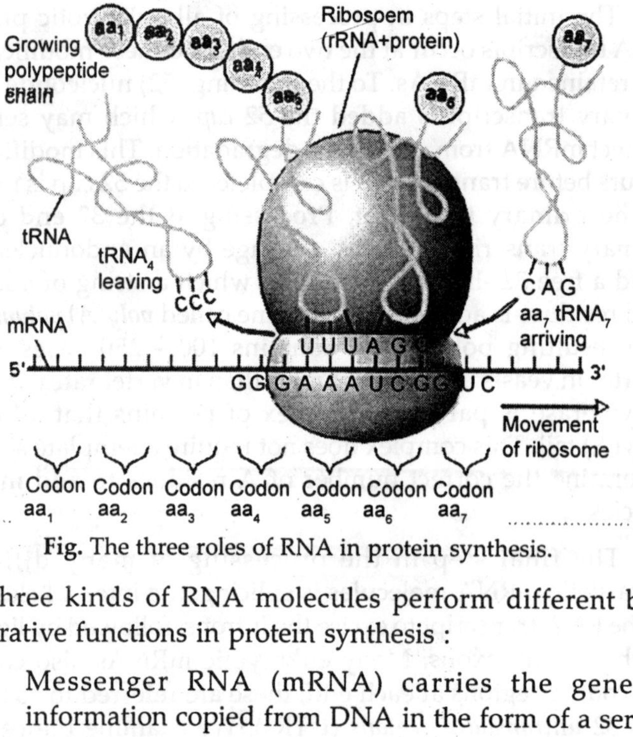

Fig. The three roles of RNA in protein synthesis.

Three kinds of RNA molecules perform different but cooperative functions in protein synthesis :

1. Messenger RNA (mRNA) carries the genetic information copied from DNA in the form of a series of three-base code "words," each of which specifies a particular amino acid.
2. Transfer RNA (tRNA) is the key to deciphering the code words in mRNA. Each type of amino acid has its own type of tRNA, which binds it and carries it to the growing end of a polypeptide chain if the next code word on mRNA calls for it. The correct tRNA with its attached amino acid is selected at each step

because each specific tRNA molecule contains a three-base sequence that can base-pair with its complementary code word in the mRNA.

3. Ribosomal RNA (rRNA) associates with a set of proteins to form ribosomes. These complex structures, which physically move along an mRNA molecule, catalyze the assembly of amino acids into protein chains. They also bind tRNAs and various accessory molecules necessary for protein synthesis. Ribosomes are composed of a large and small subunit, each of which contains its own rRNA molecule or molecules.

Translation is the whole process by which the base sequence of an mRNA is used to order and to join the amino acids in a protein. The three types of RNA participate in this essential protein-synthesizing pathway in all cells; in fact, the development of the three distinct functions of RNA was probably the molecular key to the origin of life. How each RNA carries out its specific task is discussed in this section, while the biochemical events in protein synthesis and the required protein factors are described in the final section of the chapter.

MESSENGER RNA CARRIES INFORMATION FROM DNA IN A THREE-LETTER GENETIC CODE

RNA contains ribonucleotides of adenine, cytidine, guanine, and uracil; DNA contains deoxyribonucleotides of adenine, cytidine, guanine, and thymine. Because 4 nucleotides, taken individually, could represent only 4 of the 20 possible amino acids in coding the linear arrangement in proteins, a *group* of nucleotides is required to represent each amino acid. The code employed must be capable of specifying at least 20 words (i.e., amino acids).

If two nucleotides were used to code for one amino acid, then only 16 (or 4^2) different code words could be formed, which would be an insufficient number. However, if a group of three nucleotides is used for each code word, then 64 (or 4^3) code words can be formed. Any code using groups of three or more nucleotides will have more than enough units to encode 20 amino acids. Many such coding systems are mathematically possible. However, the actual genetic code used by cells is a *triplet* code, with every three nucleotides being "read" from a specified starting point in the mRNA. Each triplet is called a codon. Of the 64 possible codons in the genetic code, 61 specify individual amino acids and three are stop codons. Only two — methionine and tryptophan — have a single codon; at the other extreme, leucine, serine, and arginine are each specified by six different codons. The different codons for a given amino acid are said to be *synonymous*. The code itself is termed *degenerate,* which means that it contains redundancies.

Synthesis of all protein chains in prokaryotic and eukaryotic cells begins with the amino acid methionine. In most mRNAs, the start (initiator) codon specifying this aminoterminal methionine is AUG. In a few bacterial mRNAs, GUG is used as the initiator codon, and CUG occasionally is used as an initiator codon for methionine in eukaryotes. The three codons UAA, UGA, and UAG do not specify amino acids but constitute stop (*terminator*) signals that mark the carboxyl terminus of protein chains in almost all cells. The sequence of codons that runs from a specific start site to a terminating codon is called a reading frame. This precise linear array of ribonucleotides in groups of three in mRNA specifies the precise linear sequence of amino acids in a protein and also signals where synthesis of the protein chain starts and stops.

Because the genetic code is a commaless, overlapping triplet code, a particular mRNA theoretically could be translated in three different reading frames. Indeed some mRNAs have been shown to contain overlapping information that can be translated in different reading frames, yielding different polypeptides. The vast majority of mRNAs, however, can be read in only one frame because stop codons encountered in the other two possible reading frames terminate translation before a functional protein is produced. Another unusual coding arrangement occurs be- cause of *frameshifting*. In this case the protein-synthesizing machinery may read four nucleotides as one amino acid and then continue reading triplets, or it may back up one base and read all succeeding triplets in the new frame until termination of the chain occurs. These frameshifts are not common events, but a few dozen such instances are known.

The meaning of each codon is the same in most known organisms — a strong argument that life on earth evolved only once. Recently the genetic code has been found to differ for a few codons in many mitochondria, in ciliated protozoans, and in *Acetabularia,* a single-celled plant. As shown most of these changes involve reading of normal stop codons as amino acids, not an exchange of one amino acid for another. It is now thought that these exceptions to the general code are later evolutionary developments; that is, at no single time was the code immutably fixed, although massive changes were not tolerated once a general code began to function early in evolution.

EXPERIMENTS WITH SYNTHETIC MRNAS AND TRINUCLEOTIDES BROKE THE GENETIC CODE

Having described the genetic code, we briefly recount how it was deciphered — one of the great triumphs of modern biochemistry. The underlying experimental work was carried out largely with cell-free bacterial extracts containing all the necessary components for protein synthesis except mRNA (i.e., tRNAs, ribosomes, amino acids, and the energy-rich nucleotides ATP and GTP).

Initially, researchers added synthetic mRNAs containing a single type of nucleotide to such extracts and then determined the amino acid incorporated into the polypeptide that was formed. In the first successful experiment, synthetic mRNA composed only of U residues [poly(U)] yielded polypeptides made up only of phenylalanine. Thus it was concluded that a codon for phenylalanine consisted entirely of U's. Likewise, experiments with poly(C) and poly(A) showed that a codon for proline contained only C's and a codon for lysine only A's. [Poly(G) did not work in this type of experiment because it assumes an unusable stacked structure that is not translated well.] Next, synthetic mRNAs composed of alternating bases were used. The results of these experiments not only revealed more codons but also demonstrated that codons are three bases long. The example of this approach led to identification of ACA as the codon for threonine and CAC for histidine. Similar experiments with many such mixed polynucleotides revealed a substantial part of the genetic code.

The entire genetic code was finally worked out by a second type of experiment conducted by Marshall Nirenberg and his collaborators. In this approach, all the possible trinucleotides were tested for their ability to attract tRNAs attached to the 20 different amino acids found in natural proteins. In all, 61 of the 64 possible trinucleotides were found to code for a specific amino acid; the trinucleotides UAA, UGA, and UAG did not encode amino acids.

Although synthetic mRNAs were useful in deciphering the genetic code, in vitro protein synthesis from these mRNAs is very inefficient and yields polypeptides of variable size. Successful in vitro synthesis of a naturally occurring protein was achieved first when mRNA from bacteriophage F2 (a virus) was added to bacterial extracts, leading to formation of the coat, or capsid, protein (the "packaging" protein that covers the virus particle). Studies with such natural mRNAs established that AUG encodes methionine at the start of almost all proteins and is required for efficient initiation of protein synthesis, while the three trinucleotides (UAA, UGA, and UAG) that do not encode any amino acid act as stop codons, necessary for precise termination of synthesis.

THE FOLDED STRUCTURE OF TRNA PROMOTES ITS DECODING FUNCTIONS

The next step in understanding the flow of genetic information from DNA to protein was to determine how the nucleotide sequence of mRNA is converted into the amino acid sequence of protein. This decoding process requires two types of adapter molecules: tRNAs and enzymes called *aminoacyl-*

tRNA synthetases. First we describe the role of tRNAs in decoding mRNA codons, and then examine how synthetases recognize tRNAs.

All tRNAs have two functions: to be chemically linked to a particular amino acid and to base-pair with a codon in mRNA so that the amino acid can be added to a growing peptide chain. Each tRNA molecule is recognized by one and only one of the 20 aminoacyl-tRNA synthetases. Likewise, each of these enzymes links one and only one of the 20 amino acids to a particular tRNA, forming an aminoacyl-tRNA. Once its correct amino acid is attached, a tRNA then recognizes a codon in mRNA, thereby delivering its amino acid to the growing polypeptide.

As studies on tRNA proceeded, 30 – 40 different tRNAs were identified in bacterial cells and as many as 50 – 100 in animal and plant cells. Thus the number of tRNAs in most cells is more than the number of amino acids found in proteins (20) and also differs from the number of codons in the genetic code (61). Consequently, many amino acids have more than one tRNA to which they can attach (explaining how there can be more tRNAs than amino acids); in addition, many tRNAs can attach to more than one codon (explaining how there can be more codons than tRNAs). As noted previously, most amino acids are encoded by more than one codon, requiring some tRNAs to recognize more than one codon.

The function of tRNA molecules, which are 70 – 80 nucleotides long, depends on their precise three-dimensional structures. In solution, all tRNA molecules fold into a similar stem-loop arrangement that resembles a cloverleaf when drawn in two dimensions. The four stems are short double helices stabilized by Watson-Crick base pairing; three of the four stems have loops containing seven or eight bases at their ends, while the remaining, unlooped stem contains the free 32 and 52 ends of the chain. Three nucleotides termed the anticodon, located at the centre of one loop, can form base pairs with the three complementary nucleotides forming a codon in mRNA. As discussed later, specific aminoacyl-tRNA synthetases recognize the surface structure of each tRNA for a specific amino acid and covalently attach the proper amino acid to the unlooped *amino acid acceptor stem*. The 32 end of all tRNAs has the sequence CCA, which in most cases is added after synthesis and processing of the tRNA are complete. Viewed in three dimensions, the folded tRNA molecule has an L shape with the anticodon loop and acceptor stem forming the ends of the two arms.

Besides addition of CCA at the 32 terminus after a tRNA molecule is synthesized, several of its nucleic acid bases typically are modified. For example, most tRNAs are synthesized with a four-base sequence of UUCG near the middle of the molecule. The first uridylate is methylated to become a thymidylate; the second is rearranged into a pseudouridylate (abbreviated Y), in which the ribose is attached to carbon 5 instead of to nitrogen 1 of the uracil. These modifications produce a characteristic TYCG loop in an unpaired region at approximately the same position in nearly all tRNAs.

NONSTANDARD BASE PAIRING OFTEN OCCURS BETWEEN CODONS AND ANTICODONS

If perfect Watson-Crick base pairing were demanded between codons and anticodons, cells would have to contain exactly 61 different tRNA species, one for each codon that specifies an amino acid. As noted above, however, many cells contain fewer than 61 tRNAs. The explanation for the smaller number lies in the capability of a single tRNA anticodon to recognize more than one, but not necessarily every, codon corresponding to a given amino acid. This broader recognition can occur because of nonstandard pairing between bases in the so-called "wobble" position: the third base in a mRNA codon and the corresponding first base in its tRNA anticodon. Although the first and second bases of a codon form standard Watson-Crick base pairs with the third and second bases of the corresponding anticodon, four nonstandard interactions can occur between bases in the wobble position. Particularly important is the G·U base pair, which structurally fits almost as well as the standard G·C pair. Thus, a given anticodon in tRNA with G in the first (wobble) position can base-pair with the two corresponding codons that have either pyrimidine (C or U) in the third position. For example, the phenylalanine codons UUU and UUC (52 → 32) are both recognized by the tRNA that has GAA (52 → 32) as the anticodon. In fact, any two codons of the type NNPyr (N = any base; Pyr = pyrimidine) encode a single amino acid and are decoded by a single tRNA with G in the first (wobble) position of the anticodon.

Although adenine rarely is found in the anticodon wobble position, many tRNAs in plants and animals contain *inosine* (I), a deaminated product of adenine, at this position. Inosine can form nonstandard base pairs with A, C, and U. A tRNA with inosine in the wobble position thus can recognize the corresponding mRNA codons with A, C, or U in the third (wobble) position. For this reason, inosine-containing tRNAs are heavily employed in translation of the synonymous codons that specify a single amino acid. For example, four of the six codons for leucine have a 32 A, C, or U; these four codons are all recognized by the same tRNA (32 -GAI-52), which has inosine in the wobble position of the anticodon (and thus recognizes CUA, CUC, and CUU), and uses a G·U pair in position 1 to recognize the UUA codon.

Aminoacyl-tRNA Synthetases Activate Amino Acids by Linking Them to tRNAs

Recognition of the codon or codons specifying a given amino acid by a particular tRNA is actually the second step in decoding the genetic message. The first step, attachment of the appropriate amino acid to a tRNA, is catalyzed by a specific aminoacyl-tRNA synthetase. Each of the 20 different synthetases recognizes *one* amino acid and *all* its compatible, or *cognate,* tRNAs. These coupling enzymes link an amino acid to the free 22 or 32 hydroxyl of the adenosine at the 32 terminus of tRNA molecules by a two-step ATP-requiring reaction. About half the aminoacyl-tRNA synthetases transfer the aminoacyl group to the 22 hydroxyl of the terminal adenosine (class I), and about half to the 32 hydroxyl (class II). In this reaction, the amino acid is linked to the tRNA by a high-energy bond and thus is said to be *activated*. The energy of this bond subsequently drives the formation of peptide bonds between adjacent amino acids in a growing polypeptide chain. The equilibrium of the aminoacylation reaction is driven further toward activation of the amino acid by hydrolysis of the high-energy phosphoanhydride bond in pyrophosphate. The overall reaction is

$$\text{Amino acid} + \text{ATP} + \text{tRNA} \xrightarrow{\textit{enzyme}} \text{aminoacyl-tRNA} + \text{AMP} + 2\text{P}i$$

The amino acid sequences of the aminoacyl-tRNA synthetases (ARSs) from many organisms are now known, and the three-dimensional structures of over a dozen enzymes of both classes have been solved. Each of these enzymes has a rather precise binding site for ATP (GTP is not admitted and CTP and UTP are too small) and binding pockets for its specific amino acid. Class I and class II enzymes bind to opposite faces of the incoming tRNAs. The binding surfaces of class I enzymes tend to be somewhat complementary to those of class II enzymes. These different binding surfaces and the consequent alignment of bound tRNAs probably account in part for the difference in the hydroxyl group to which the aminoacyl group is transferred. Because some amino acids are so similar structurally, aminoacyl-tRNA synthetases sometimes make mistakes. These are corrected, however, by the enzymes themselves, which check the fit in the binding pockets and facilitate deacylation of any misacylated tRNAs. This crucial function helps guarantee that a tRNA delivers the correct amino acid to the protein-synthesizing machinery.

EACH TRNA MOLECULE IS RECOGNIZED BY A SPECIFIC AMINOACYL-TRNA SYNTHETASE

The ability of aminoacyl-tRNA synthetases to recognize their correct cognate tRNAs is just as important to the accurate translation of the genetic code as codon-anticodon pairing. Once a tRNA is loaded with an amino acid, codon-anticodon pairing directs the tRNA into the proper ribosome site; if the wrong amino acid is attached to the tRNA, an error in protein synthesis results.

As noted already, each aminoacyl-tRNA synthetase can aminoacylate all the different tRNAs whose anticodons correspond to the same amino acid. Therefore, all these cognate tRNAs must have a similar binding site, or "identity element," that is recognized by the synthetase. One approach for studying the identity elements in tRNAs that are recognized by aminoacyl-tRNA synthetases is to produce synthetic genes that encode tRNAs with normal and various mutant sequences by techniques. The normal and mutant tRNAs produced from such synthetic genes then can be tested for their ability to bind purified synthetases.

Very probably no single structure.or sequence completely determines a specific tRNA identity. However, some important structural features of several *E. coli* tRNAs that allow their cognate synthetases to recognize them are known. Perhaps the most logical identity element in a tRNA molecule is the anticodon itself. Experiments in which the anticodons of methionine tRNA ($tRNA^{Met}$) and valine tRNA ($tRNA^{Val}$) were interchanged showed that the anticodon is of major importance in determining the identity of these two tRNAs. In addition, x-ray crystallographic analysis of the complex between glutamine aminoacyl-tRNA synthetase (GlnRS) and glutamine tRNA ($tRNA^{Gln}$) showed that each of the anticodon bases neatly fits into a separate, specific "pocket" in the three-dimensional structure of GlnRS. Thus this synthetase specifically recognizes the correct anticodon.

However, the anticodon may not be the principal *identity element* in other tRNAs. Identity elements are found in several regions, particularly the end of the acceptor arm. A simple case is presented by $tRNA^{Ala}$: a single G·U base pair (G3·U70) in the acceptor stem is necessary and sufficient for recognition of this tRNA by its cognate aminoacyl-tRNA synthetase. Solution of the three-dimensional structure of additional complexes between aminoacyl-tRNA synthetases and their cognate tRNAs should provide a clear understanding of the rules governing the recognition of tRNAs by specific synthetases.

RIBOSOMES ARE PROTEIN-SYNTHESIZING MACHINES

If the many components that participate in translating mRNA had to interact in free solution, the likelihood of simultaneous collisions occurring would be so low that the rate of amino acid polymerization would be very slow. The efficiency of translation is greatly increased by the binding of the mRNA and the individual aminoacyl-tRNAs to the most abundant RNA-protein complex in the cell — the ribosome. This two-part machine directs the elongation of a polypeptide at a rate of three to five amino acids added per second. Small proteins of 100 – 200 amino acids are therefore made in a minute or less. On the other hand, it takes 2 to 3 hours to make the largest known protein, titin, which is found in muscle and contains 30,000 amino acid residues. The machine that accomplishes this task must be precise and persistent.

With the aid of the electron microscope, ribosomes were first discovered as discrete, rounded structures prominent in animal tissues secreting large amounts of protein; initially, however, they were not known to play a role in protein synthesis. Once reasonably pure ribosome preparations were obtained, radiolabeling experiments showed that radioactive amino acids first were incorporated into growing polypeptide chains associated with ribosomes before appearing in finished chains.

A ribosome is composed of several different ribosomal RNA (rRNA) molecules and more than 50 proteins, organized into a large subunit and a small subunit. The proteins in the two subunits differ, as do the molecules of rRNA. The small ribosomal subunit contains a single rRNA molecule, referred to as *small rRNA;* the large subunit contains a molecule of *large rRNA* and one molecule each of two much smaller rRNAs in eukaryotes. The ribosomal subunits and the rRNA molecules are commonly designated in svedbergs (S), a measure of the sedimentation rate of suspended particles centrifuged under standard conditions. The lengths of the rRNA molecules, the quantity of proteins in each subunit, and consequently the sizes of the subunits differ in prokaryotic and eukaryotic cells. (The small and large rRNAs are about 1500 and 3000 nucleotides long in bacteria and about 1800 and 5000 nucleotides long in humans.) Perhaps of more interest than these differences are the great structural and functional similarities among ribosomes from all species. This consistency is another reflection of the common evolutionary origin of the most basic constituents of living cells.

The sequences of the small and large rRNAs from several thousand organisms are now known. Although the primary nucleotide sequences of these rRNAs vary considerably, the same parts of each type of rRNA theoretically can form base-paired stem-loops, generating a similar threedimensional structure for each rRNA in all organisms. Evidence that such stem-loops occur in rRNA was obtained by treating rRNA with chemical agents that cross-link paired bases; the samples then were digested with enzymes that destroy single-stranded rRNA, but not any cross-linked (base-paired) regions. Finally, the intact, cross-linked rRNA that remained was collected and sequenced, thus identifying the stem-loops in the original rRNA. Experiments of this type have located about 45 stem-loops at similar positions in small rRNAs from many different prokaryotes and eukaryotes. An even larger number of regularly positioned stem-loops have been demonstrated in large rRNAs. All the ribosomal proteins have been identified and their sequences determined, and many have been shown to bind specific regions of rRNA. It seems clear that the fundamental protein-synthesizing machinery in all present-day cells arose only once and has been modified about a common plan during evolution.

During protein synthesis, a ribosome moves along an mRNA chain, interacting with various protein factors and tRNA and very likely even undergoes shape changes. Despite the complexity of the ribosome, great progress has been made in determining both the overall structure of bacterial ribosomes and in identifying reactive sites that bind specific proteins, mRNA, and tRNA and that participate in important steps in protein synthesis. Quite detailed models of the large and small ribosomal subunits from *E. coli* have been constructed based on cryoelectron microscope and neutron-scattering studies. These studies not only have determined the dimensions and overall shape of the ribosomal subunits, but also have localized the positions of tRNAs bound to the ribosome during protein chain elongation. Powerful chemical experiments have also helped unravel the complex interactions between proteins and RNAs. In a technique called footprinting, for example, ribosomes are treated with chemical reagents that modify single-stranded RNA unprotected by binding either to protein or to another RNA. If the total sequence of the RNA is known, then the location of the modified nucleotides can be located within the molecule. Thus the overall structure and function of ribosomes during protein synthesis is finally, after 40 years, yielding to successful experiments. How these results aid in understanding the specific steps in protein synthesis is described in the next section.

STEPWISE FORMATION OF PROTEINS ON RIBOSOMES

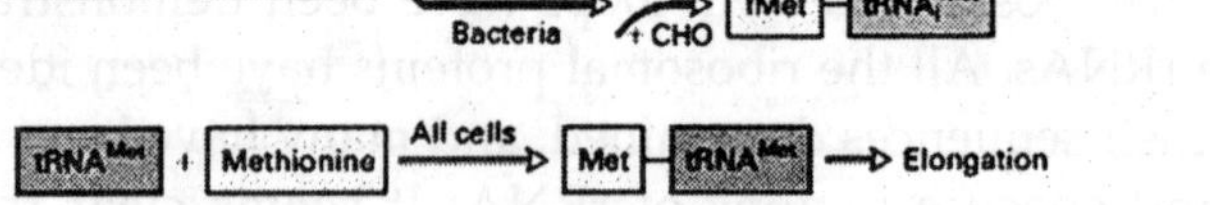

Fig Two types of methionine tRNA are found in all cells.

The previous sections have introduced the major participants in protein synthesis — mRNA, aminoacylated tRNAs, and ribosomes containing large and small rRNAs. Here, we take a detailed look at how these components are brought together to carry out the biochemical events leading to formation of proteins on ribosomes. The complex process of translating mRNA into protein can be divided into three stages — initiation, elongation, and termination — which we consider in order.

The AUG Start Codon Is Recognized by Methionyl-$tRNA_i^{Met}$

As noted earlier, the AUG codon for methionine functions as the start codon in the vast majority of mRNAs. A critical aspect of initiation is to begin protein synthesis at the start codon, thereby setting the stage for the correct in-frame translation of the entire mRNA. Both prokaryotes and eukaryotes contain two different methionine tRNAs: $tRNA_i^{Met}$ can initiate protein synthesis, and $tRNA^{Met}$ can incorporate nethionine only into a growing protein chain. The same aminoacyl-tRNA synthetase (MetRS) charges both tRNAs with methionine. But *only* Met-$tRNA_i^{Met}$ (i.e., activated methionine attached to $tRNA_i^{Met}$) can bind at the appropriate site on the small ribosomal subunit (the P site; see discussion below) to begin synthesis of a protein chain. (The regular $tRNA^{Met}$ binds only to another ribosomal site, the A site, as described later.) In bacteria, but not in archaeans or eukaryotes, the amino group of the methionine of Met-$tRNA_i^{Met}$ is modified by addition of a formyl group and is sometimes designated fMet-$tRNA_i^{Met}$. However, Met-$tRNA_i^{Met}$ is commonly used to designate the initiator tRNA in all cells.

During the initial stage of protein synthesis in all cells, a ribosome assembles, complexed with a mRNA and an activated initiator tRNA, which is correctly positioned at the start codon. Because the details of initiation and the mechanism for locating the translation start site differ in bacteria and eukaryotes, we discuss the two systems separately.

Bacterial Initiation of Protein Synthesis Begins Near a Shine-Dalgarno Sequence in mRNA

To begin the assembly of a bacterial translation complex, sequential interactions occur between specific proteins referred to as initiation factors (IFs) and the small (30S) ribosomal subunit. The resulting preinitiation complex together with fMet-$tRNA_i^{Met}$ then binds to the mRNA at a specific site usually located quite near the AUG initiation codon. This binding, assisted by IF1 and IF3, yields the *30S initiation complex*.

In most bacteria, the small ribosomal subunit identifies start codons through interaction of the small (16S) rRNA with an eight-nucleotide sequence in mRNA called the *Shine-Dalgarno sequence* after its discoverers. This sequence, located near the AUG start codon, base-pairs to a sequence at or very near the 32 end of 16S rRNA, thereby binding the mRNA and small ribosomal subunit to each other. Although the Shine-Dalgarno sequence varies some at different initiation sites, on average six out of eight nucleotides are complementary to the 32 -end sequence of 16S rRNA. Thus bacterial rRNA plays a direct role in recruiting the small ribosomal subunit to a translation start site on the mRNA. The long mRNAs produced from bacterial operons contain multiple *internal* Shine-Dalgarno sequences located near the start sites for each of the encoded proteins. Because a small ribosomal subunit can bind to each Shine-Dalgarno sequence, synthesis of the different proteins can occur independently and simultaneously from the multiple start sites on these long mRNAs.

Assembly of a complete *70S initiation complex* is achieved by the recruitment of the large ribosomal subunit, an energy-requiring step that is powered by hydrolysis of GTP bound to IF2; in the process IF1, IF2-GDP, and P_i are discharged. In the completed initiation complex, the charged initiator tRNA is positioned at the *P site* on the ribosome. As discussed later, during elongation the growing peptide chain remains attached to the tRNA located at this site.

EUKARYOTIC INITIATION OF PROTEIN SYNTHESIS OCCURS AT THE 5′ END AND INTERNAL SITES IN MRNA

Initiation of protein synthesis in eukaryotic cells, as in bacteria, begins with formation of a preinitiation complex prior to mRNA binding. However, the process differs significantly from initiation in bacteria. Two eukaryotic factors, eIF3 (a large multimeric protein with about eight subunits) and eIF6, serve to keep the ribosomal subunits apart after they have previously finished synthesizing a protein chain so that they can participate again in starting a new chain. To assemble a eukaryotic preinitiation complex, an active "ternary complex" of eIF2 bound to a GTP molecule and Met-$tRNA_i^{Met}$ associates with a small (40S) ribosomal subunit complexed with two other factors, eIF3 and eIF1A, which stabilize binding of the ternary complex. Cells can regulate protein synthesis by phosphorylating a serine residue on the eIF2 bound to GDP; this complex is then unable to bind Met-$tRNA_i^{Met}$, thus inhibiting protein synthesis.

Recognition of the translation start site by the small ribosomal subunit in the eukaryotic preinitiation complex requires different types of factors than in prokaryotic cells. Most eukaryotic mRNAs have a single start site near the 52 capped end of the mRNA. Interaction of the small ribosomal subunit with the methylated 52 -cap structure present on all eukaryotic mRNAs requires a set of proteins collectively called *eIF4*. After the methylated cap structure is recognized by eIF4F, any secondary structure at the 52 end is removed by an associated helicase activity. The bound preinitiation complex, which contains the eIF3 group of proteins, then probably slides along the mRNA, most often stopping at the first AUG. However, selection of the initiating AUG is facilitated by specific surrounding sequences called *Kozak sequences* (for Marilyn Kozak, who defined them):

mRNA 5'–ACCAUGG–

The A preceding the AUG seems to be the most important nucleotide affecting initiation efficiency. During the selection of the starting AUG, there is a proven interaction between the eIF3 complex, the 40S ribosomal subunit, and the eIF4G component of the eIF4 complex. Scanning of the mRNA by the preinitiation complex eventually yields a *40S initiation complex* in which Met-$tRNA_i^{Met}$ is correctly positioned at the translation start site.

Once the small ribosomal subunit with its bound Met-tRNA$_i^{Met}$ is correctly positioned at the start codon, union with the large (60S) ribosomal subunit completes formation of the *80S initiation complex.*

Initiation of translation of most mRNAs by the eukaryotic protein-synthesizing machinery begins near the 52 capped end as just described. However, some viral mRNAs, which are translated by the host-cell machinery in infected eukaryotic cells, lack a 52 cap; translation of these mRNAs is initiated at *internal ribosome entry sites* (IRESs). It is now known that some cellular mRNAs also can be translated beginning at IRESs. Many of the same protein factors that assist in ribosome scanning from a 52 cap are required for locating an internal AUG start codon, but exactly how an IRES is recognized is less clear.

DURING CHAIN ELONGATION EACH INCOMING AMINOACYL-TRNA MOVES THROUGH THREE RIBOSOMAL SITES

The correctly positioned bacterial 70S or eukaryotic 80S ribosome – Met-tRNA$_i^{Met}$ complex is now ready to begin the task of stepwise addition of amino acids by the in-frame translation of the mRNA. As is the case with initiation, a set of special proteins, termed elongation factors (EFs), are required to carry out the process. The key steps in elongation are entry of each succeeding aminoacyl-tRNA, formation of a peptide bond, and the movement, or *translocation,* of the ribosome with respect to the mRNA. For example, the second aminoacyl-tRNA is brought into the ribosome as a ternary complex in association with an EF1 α-GTP in eukaryotes (EF-Tu – GTP in bacteria) and becomes bound to the A site on the ribosome. Recall that the initiating Met-tRNA$_i^{Met}$ is bound at the P site and base-paired with the AUG start codon. If the anticodon of the incoming (second) aminoacyl-tRNA correctly matches the second codon of the mRNA, a tight binding ensues at the A site. If this second codon does not match the incoming aminoacyl-tRNA, it diffuses away. The choice of the correct aminoacyl-tRNA and its tight binding at the A site requires energy that is supplied by hydrolysis of the EF1 α-GTP (or EF-Tu – GTP) complex.

With the initiating Met-$tRNA_i^{Met}$ at the P site and the second aminoacyl-tRNA tightly bound at the A site, the α amino group of the second amino acid reacts with the "activated" (aminoacylated) methionine on the initiator tRNA, forming a peptide bond. A revolutionary piece of biochemistry has provided insight about this critical *peptidyltransferase reaction,* which actually effects the transfer of the growing peptide chain on the peptidyl-tRNA at the P site to the activated amino acid on the incoming aminoacyl-tRNA. In bacterial protein synthesis, the 23S rRNA in the large ribosomal subunit itself may carry out the peptidyltransferase function. Supporting this catalytic role for 23S rRNA is the finding that when the vast majority of the protein is carefully removed from the large ribosomal subunit and the remaining rRNA, or even a pure segment of 23S rRNA, is mixed with analogs of aminoacylated-tRNA and peptidyl-tRNA, the peptidyltransferase reaction still occurs.

During the process of peptide synthesis and tRNA site changes, the ribosome is moved along the mRNA a distance equal to one codon with the addition of each amino acid. This translocation step is catalyzed by eukaryotic EF2-GTP (bacterial EF-G – GTP), which is hydrolyzed to provide the required energy. We can see that after peptide linkage $tRNA_i^{Met}$, now without its activated methionine, is moved to an exit (E) site on the ribosome and is soon discharged. Concurrently, another ternary complex, carrying the next amino acid to be added, enters the ribosome, and the cycle continues.

PROTEIN SYNTHESIS IS TERMINATED BY RELEASE FACTORS WHEN A STOP CODON IS REACHED

The final phase of protein synthesis, like initiation and elongation, requires highly specific molecular signals that decide the fate of the mRNA-ribosome-tRNA-peptidyl complex. Recent research has uncovered two types of specific termination (release) factors in bacteria and their eukaryotic counterparts. Bacterial RF1 and RF2 (and eukaryotic eRF1), whose shape is thought to be similar to that of tRNAs ("molecular mimicry"), apparently act by recognizing the stop codons themselves. RF1 recognizes UAG, and RF2 recognizes UGA; both these factors recognize UAA. Like some of the initiation factors and the elongation factors discussed previously, the third release factor, RF3 (and its eukaryotic counterpart), is a GTP-binding protein. RF3 acts in concert with the codon-recognizing factors to promote cleavage of the peptidyl-tRNA, thus releasing the completed protein chain.

Folding of a newly released protein into its native three-dimensional conformation is facilitated by other proteins called *chaperones*. Additional release factors then promote dissociation of the ribosome, freeing the subunits, mRNA, and terminal tRNA for another round of protein synthesis.

Simultaneous Translation by Multiple Ribosomes and Their Rapid Recycling Increase the Efficiency of Protein Synthesis

As noted earlier, translation of a single eukaryotic mRNA molecule to yield a typical-sized protein takes 30 – 60 seconds. Two phenomena significantly increase the overall rate at which cells can synthesize a protein: the simultaneous translation of a single mRNA molecule by multiple ribosomes and rapid recycling of ribosomal subunits after they disengage from the 32 end of a mRNA. Simultaneous translation of an mRNA by multiple ribosomes is readily observable in electron micrographs and by sedimentation analysis, revealing mRNA attached to multiple ribosomes bearing nascent growing peptide chains. These structures are referred to as *polyribosomes* or *polysomes*. However, the second phenomenon — rapid recycling of ribosomal subunits — was not discovered so easily.

Polysomes in some tissues were seen in electron micrographs to be circular. It was not until recently that studies with yeast cells explained the circular shape of polyribosomes and suggested the mode by which ribosomes recycle efficiently.

Researchers initially found that the poly(A) tail at the 32 end of eukaryotic mRNA stimulates in vitro translation by yeast protein-synthesizing components. This observation at first seems puzzling, since translation is begun upstream (usually near the 52 end) of the mRNA, at a great distance from the poly(A) tail. A cytosolic protein found in all eukaryotic cells, *poly(A)-binding protein I* (PABI), proved literally to be a missing link in explaining the relationship of the poly(A) tail to translation. In earlier studies, yeast cells that had mutations in the gene encoding PABI were found to translate mRNA poorly. The role of PABI in improving efficiency of protein synthesis was demonstrated by the recent finding that PABI and two components of yeast eIF4 (4G and 4E) can interact on mRNA to circularize the molecule. By linking the poly(A) tail to the 52 end, PABI positions recently disengaged ribosomal subunits near the start site for protein synthesis, thus enhancing their reentry into the process and improving efficiency. Thus in many eukaryotic cells, ribosomal subunits may follow the circular pathway providing another example of the merger of form and function.

Chapter 5

Biomembranes and the Subcellular Organization of Eukaryotic Cells

Recognition that modern-day organisms fall into three distinct evolutionary lineages came from comparisons of the nucleotide sequences of genes common to all organisms and of ribosomal RNAs, particularly the RNA species found in the small ribosomal subunit. Prokaryotes, which lack a defined nucleus and have a simple subcellular organization, form two of the lineages — the bacteria and archaea. The third lineage is the eukaryotes (eukarya), whose cells have a membrane-limited nucleus containing most of the cellular DNA, numerous specialized organelles, and a complex cytoskeleton. Despite the differences in cellular organization between prokaryotes and eukaryotes, all cells share certain structural features and carry out DNA replication, protein synthesis, production of ATP, and many other complicated metabolic events in basically the same way. At the molecular level there is a close relationship between protein structure and function. Likewise, at the cellular level, function and structure are intimately related. Thus this chapter describes the basic structural components of cells and the general functions of each, focusing on eukaryotic cells.

We begin by describing the capabilities and applications of various techniques of microscopy used to visualize cells and subcellular structures. We then consider the basic methods for purifying biological membranes and subcellular structures.

The purified preparations produced by cell fractionation permit detailed studies on the structure and function of biomembranes and other subcellular structures. Following this overview of experimental techniques, we continue the discussion of biomembranes begun in earlier chapters. The surface membrane found on all cells and the membranes that line eukaryotic organelles all have the same basic architecture — a phospholipid bilayer; the unique function of each type of membrane is determined primarily by the proteins it contains. The final section briefly describes the structure and function of the main internal organelles and fibers of eukaryotic cells, as well as the extracellular substances that surround cells and give them shape and strength. This review sets the stage for later discussions of cell function at the molecular level.

MICROSCOPY AND CELL ARCHITECTURE

The modern, detailed understanding of cell architecture is based on several types of microscopy. Because there is no one "correct" view of a cell, it is essential to understand the characteristics of the key cell-viewing techniques, the types of images they produce, and their limitations.

Schleiden and Schwann, using a primitive light microscope, first described individual cells as the fundamental unit of life, and light microscopy has continued to play a major role in biological research. The development of electron microscopes greatly extended the ability to resolve subcellular particles and has yielded much new information on the organization of plant and animal tissues. The nature of the images depends on the type of light or electron microscope employed and on the way in which the cell or tissue has been prepared. Each technique is designed to emphasize particular structural features of the cell. The epithelial cell lining the small intestine, appears when viewed by three different microscopic techniques.

In this section, we focus on the most common application of light and electron microscopy — to visualize fixed, killed cells. Although this approach reveals much information, a critical question about such results is how true to life is the image of a biological specimen that has been fixed, stained, and dehydrated before examination? Thus we also consider some of the refinements that allow microscopy of unaltered or less altered specimens.

Light Microscopy Can Distinguish Objects Separated by 0.2 μm or More

The *compound microscope,* the most common microscope in use today, contains several lenses that magnify the image of a specimen under study. The total magnification is a product of the magnification of the individual lenses: if the *objective lens* magnifies 100-fold (a 100X lens, the maximum usually employed) and the *eyepiece* magnifies 10-fold, the final magnification recorded by the human eye or on film will be 1000-fold.

However, the most important property of any microscope is not its magnification but its resolving power, or resolution — its ability to distinguish between two very closely positioned objects. Merely enlarging the image of a specimen accomplishes nothing if the image is blurry. The resolution of a microscope lens is numerically equivalent to D, the minimum distance between two distinguishable objects; the smaller the value of D, the better the resolution. D depends on three parameters, all of which must be considered in order to achieve the best possible resolution: the *angular aperture,* α, or half-angle of the cone of light entering the objective lens from the specimen; the *refractive index,* N, of the air or fluid medium between the specimen and the objective lens; and the *wavelength,* λ, of incident light: $D = (0.61\lambda) \div (N \times \sin \alpha)$. Decreasing the value of λ or increasing either N or α will decrease the value of D and thus improve the resolution. Note that the magnification is not part of this equation.

The angular aperture, α, depends on the width of the objective lens and its distance from the specimen. Moving the objective lens closer to the specimen increases the angle α and thus $\sin \alpha$, and therefore reduces D (i.e., increases the resolution). Intuitively, one can recognize that increasing α

allows a greater fraction of the light emanating from the specimen to enter the objective lens. The refractive index N is a measure of the degree to which a medium bends a light ray that passes through it; the refractive index of air is defined as 1.0. Use of immersion oil, which has a refractive index of 1.5, is a simple way to reduce D by 33 percent. An intuitive explanation for this improvement is that a medium with a higher refractive index than air, if placed between the specimen and the objective lens, will "bend" more of the light emanating from the specimen such that it goes into the lens. Finally, the shorter the wavelength of incident light, the lower will be the value of D and the better the resolution.

Due to limitations on the values of α, λ, and N, the *limit of resolution* of a light microscope using visible light is about 0.2 μm (200 nm). No matter how many times the image is magnified, the microscope can never resolve objects that are less than ≈0.2 μm apart or reveal details smaller than ≈0.2 μm in size. This is true because the maximum angular aperture for the best objective lenses is 70° (sin 70° = 0.94). With the visible light of shortest wavelength (blue, $\lambda = 450$ nm) and with an immersion oil ($N = 1.5$) above the sample, then or about 0.2 μm.

$$D = \frac{0.61 \times 450nm}{1.5 \times 0.94} = 194nm$$

Despite this limit of resolution, the light microscope can be used to track the location of a small bead of known size to a precision of only a few nanometers! If we know the precise size and shape of an object — say, a 5-nm sphere of gold — and if we use a video camera to record the microscopic image as a digital image, then a computer can calculate the position of the *centre* of the object to within a few nanometers. This technique has been used, to nanometer resolution, for tracking the movement of gold particles attached via antibodies to specific proteins on the surface of living cells.

SAMPLES FOR LIGHT MICROSCOPY USUALLY ARE FIXED, SECTIONED, AND STAINED

Specimens for light microscopy are commonly fixed with

a solution containing alcohol or formaldehyde, compounds that denature most proteins and nucleic acids. Formaldehyde also cross-links amino groups on adjacent molecules; these covalent bonds stabilize protein-protein and protein – nucleic acid interactions and render the molecules insoluble and stable for subsequent procedures. Usually the sample is then embedded in paraffin or plastic and cut into thin sections of one or a few micrometers thick. Alternatively, the sample can be frozen without prior fixation and then sectioned; this avoids the denaturation of enzymes by fixatives such as formaldehyde.

Since the resolution of the light microscope is ≈0.2 μm and mitochondria and chloroplasts are ≈1 μm long (about the size of bacteria), theoretically one should be able to see these organelles. However, most cellular constituents are not coloured and absorb about the same degree of visible light, so that they are hard to distinguish under a light microscope unless the specimen is stained. Thus the final step in preparing a specimen for light microscopy is to stain it, in order to visualize the main structural features of the cell or tissue. Many chemical stains bind to molecules that have specific features. For example, *hematoxylin* binds to basic amino acids (lysine and arginine) on many different kinds of proteins, whereas *eosin* binds to acidic molecules (such as DNA, and aspartate and glutamate side chains). Because of their different binding properties, these dyes stain various cell types sufficiently differently that they are distinguishable visually. Two other common dyes are *benzidine,* which binds to heme-containing proteins and nucleic acids, and *fuchsin,* which binds to DNA and is used in Fuelgen staining.

If an enzyme catalyzes a reaction that produces a coloured or otherwise visible precipitate from a colourless precursor, the enzyme may be detected in cell sections by their coloured reaction products. This technique is called *cytochemical staining.*

FLUORESCENCE MICROSCOPY CAN LOCALIZE AND QUANTIFY SPECIFIC MOLECULES IN CELLS

Perhaps the most versatile and powerful technique for

localizing proteins within a cell by light microscopy is fluorescent staining of cells and observation in the *fluorescence microscope*. A chemical is said to be *fluorescent* if it absorbs light at one wavelength (the *excitation wavelength*) and emits light (fluoresces) at a specific and longer wavelength. Most fluorescent dyes emit visible light, but some (such as Cy5 and Cy7) emit infrared light. In modern fluorescence microscopes, only fluorescent light emitted by the sample is used to form an image; light of the exciting wavelength induces the fluorescence but is then not allowed to pass the filters placed between the objective lens and the eye or camera.

REVEALING SPECIFIC PROTEINS IN FIXED CELLS

Four very useful dyes for fluorescent staining are rhodamine and Texas red, which emit red light; Cy3, which emits orange light; and fluorescein, which emits green light. These dyes have a low, nonspecific affinity for biological molecules, but they can be chemically coupled to purified antibodies specific for almost any desired macromolecule. When a fluorescent dye – antibody complex is added to a permeabilized cell or tissue section, the complex will bind to the corresponding antigens, which then light up when illuminated by the exciting wavelength, a technique called *immunofluorescence microscopy*. By staining a specimen with two or three dyes that fluoresce at different wavelengths, multiple proteins can be localized within a cell.

REVEALING SPECIFIC PROTEINS IN LIVING CELLS

Fluorescence microscopy can also be applied to live cells. For example, purified actin may be chemically linked to a fluorescent dye. Careful biochemical studies have established that this "tagged" molecule is indistinguishable in function from its normal counterpart. If the tagged protein is *microinjected* into a cultured cell, the endogenous cellular and injected tagged actin monomers copolymerize into normal long actin fibers. This technique can also be used to study individual microtubules within a cell.

Another technique for detecting specific proteins within living cells takes advantage of *green fluorescent protein* (GFP), a naturally fluorescent protein found in the jellyfish *Aequorea victoria*. The bioluminescence of this organism, which radiates a green fluorescence, is due to GFP. This 238-aa protein contains serine, tyrosine, and glycine residues whose side chains have spontaneously reacted with one another to form a fluorescent chromophore. By recombinant DNA techniques the GFP gene can be introduced into living cultured cells or into specific cells of an entire animal. Because the introduced gene will express GFP, the cells will emit a green fluorescence when irradiated; this GFP fluorescence can be used to localize the cells within a tissue.

Alternatively, the gene for GFP can be fused to the gene for another protein of interest, producing a recombinant DNA encoding one long chimeric protein that contains the entirety of both proteins. Cells in which this recombinant DNA has been introduced will synthesize this chimeric protein, whose green fluorescence will reveal the subcellular localization of the protein.

DETERMINING THE INTRACELLULAR CONCENTRATION OF CA^{2+} AND H^{+} IONS

Changes in the cytosolic concentration of Ca^{2+} ions or pH frequently signal changes in cellular metabolism. The Ca^{2+} concentration in the cytosol of resting cells, for instance, is about 10^{-7} M. Many hormones or other stimuli cause a rise in cytosolic Ca^{2+} to 10^{-6} M; this, in turn, causes changes in cellular metabolism, such as contraction of muscle.

The fluorescent properties of certain dyes, such as *fura-2*, facilitate measurement of the concentration of free Ca^{2+} in the cytosol. This dye contains five carboxylate groups that form ester linkages with ethanol. The resulting fura-2 ester is lipophilic and can diffuse from the medium across the plasma membrane into cells. Within the cytosol, esterases hydrolyze fura-2 ester yielding fura-2, whose free carboxylate groups render the molecule nonlipophilic, so it cannot cross cellular membranes and remains in the cytosol. Each fura-2 molecule can bind a single Ca^{2+} ion but no other cellular cation, and the amount of fura-2 bound to Ca^{2+} is proportional, over a certain range, to the Ca^{2+} concentration. The fluorescence of fura-2 at one particular wavelength is enhanced when Ca^{2+} is bound, and the fluorescence is proportional to the Ca^{2+} concentration. At another wavelength the fluorescence of fura-2 is the same whether or not Ca^{2+} is bound and provides a measure of the total amount of fura-2 in the segment of the cell. By examining cells continuously in the fluorescence microscope and measuring rapid changes in the ratio of fura-2 fluorescence at these two wavelengths, one can quantify rapid changes in the fraction of fura-2 that has a bound Ca^{2+} ion and thus in the concentration of cytosolic Ca^{2+}.

The fluorescence of other dyes is sensitive to the H^+ concentration and can be used in a similar way to monitor the cytosolic pH of living cells.

CONFOCAL SCANNING AND DECONVOLUTION MICROSCOPY PROVIDE SHARPER IMAGES OF THREE-DIMENSIONAL OBJECTS

Immunofluorescence microscopy has its limitations. The fixatives employed to preserve cell architecture often destroy the *antigenicity* of a protein, that is, its ability to bind to its specific antibody. Also, the method generally gives poor results with thin cell sections, because embedding media often fluoresce themselves, obscuring the specific signal from the antibody. Moreover, in microscopy of whole cells, the fluorescent light comes from molecules above and below the plane of focus; thus the observer sees a superposition of fluorescent images from molecules at many depths in the cell, making it difficult to determine the actual three-dimensional molecular arrangement.

The *confocal scanning microscope* avoids the last problem by permitting the observer to visualize fluorescent molecules in a single plane of focus, thereby creating a vastly sharper cross-sectional image. At any instant during confocal imaging, only a single small part of a sample is illuminated with exciting light from a focused laser beam, which rapidly moves to different spots in the sample focal plane. Images from these spots are recorded by a video camera and stored in a computer, and the composite image is displayed on a computer screen.

Deconvolution microscopy is similar to confocal microscopy in that a cross-sectional image is obtained, but the two techniques differ in the details of how this image is generated. In both cases, the objective lens collects light that originates from above and below the focal plane as well as that which originates from within the focal plane. Confocal microscopes use a pinhole to exclude the out-of-focus light. In contrast, deconvolution microscopes collect all the light from several focal planes, and then mathematically reassign the out-of-focus light to its correct focal plane with the aid of a high-speed computer, a mathematical operation called *deconvolution*.

To understand how a deconvolution microscope works, consider an infinitely small fluorescent source of light, which can be approximated by a fluorescent bead smaller than the resolution of the light microscope (i.e., <0.2 μm in diameter). The emitted light radiates in all directions, and when the source is in the focal plane of the objective, it appearsas a bright point of light. When the point source is outside the focal plane of the objective, some of the light is still collected by the objective lens, and the point source appears as a halo. As the focal plane is moved farther away from the plane containing the point source, the halo becomes larger and more diffuse. Knowing exactly how the light emitted by an infinitely small fluorescent source is collected and distorted by the optics of the sample and microscope, it is possible to reconstruct an individual cross-sectional image (containing only light that originated in the focal plane of interest) from a set of images taken as the objective focal plane is moved through the plane of interest.

Cross-sectional images obtained with a deconvolution microscope may have even greater detail than those obtained with a confocal microscope. Additionally, the fluorescent labeling of the sample does not need to be as intense for deconvolution microscopy as it does for confocal microscopy, since all the light produced by a fluorescent sample is collected and analyzed by the microscope. Three-dimensional images can be obtained by a refinement known as *optical sectioning*. In this method, a computer records individual fluorescent images of planes at different depths of the sample — in effect, serial sections — and combines the stack of images into one three-dimensional image.

PHASE-CONTRAST AND NOMARSKI INTERFERENCE MICROSCOPY VISUALIZE UNSTAINED LIVING CELLS

Detailed views of transparent, live, unstained cells and tissues are obtainable with *phase-contrast microscopy* and *Nomarski interference microscopy*. Both techniques take advantage of the phenomena of refraction and diffraction of light waves. As a result, small differences in refractive index and thickness between parts of the specimen (say, between the nucleus and cytosol) or between the specimen and the surrounding medium can be converted into differences of light and dark in the final image.

The phase-contrast microscope generates an image in which the degree of darkness or brightness of a region of the sample depends on the refractive index of that region. The improved definition of subcellular structures in live, unstained cells obtained by phase-contrast microscopy compared with standard bright-field microscopy.

Nomarski, or differential, interference microscopy generates an image that looks as if the specimen is casting a shadow to one side : the "shadow" primarily represents a difference in refractive index and thickness of a specimen rather than its topography. In this technique, a prism splits an incident beam of plane-polarized light so that one part of the beam passes through one region of a specimen and the other part passes through a closely adjacent region; a second prism then reassembles the two beams. Minute differences in thickness or in the refractive index between adjacent parts of a sample are converted into a bright image (if the two beams are in phase when they recombine) or a dark one (if they are out of phase).

Phase-contrast microscopy is especially useful in examining the structure and movement of larger organelles, such as the nucleus and mitochondria, in live cultured cells. The greatest disadvantage of this technique is that it is suitable for observing only single cells or thin cell layers. Nomarski interference microscopy, in contrast, defines only the outlines of large organelles, such as the nucleus and vacuole. However, thick objects, such as the nuclei in a worm, can be observed by combining this technique with optical sectioning.

Both phase-contrast and Nomarski interference microscopy can be used in *time-lapse microscopy,* in which the same cell is photographed at regular intervals over periods of several hours. This procedure allows the observer to study cell movement, provided the microscope's stage can control the temperature of the specimen and the gas environment.

TRANSMISSION ELECTRON MICROSCOPY HAS A LIMIT OF RESOLUTION OF 0.1 NM

The fundamental principles of electron microscopy are similar to those of light microscopy; the major difference is that electromagnetic lenses, not optical lenses, focus a high velocity electron beam instead of visible light. Because electrons are absorbed by atoms in air, the entire tube between the electron source and the viewing screen is maintained under an ultrahigh vacuum.

The *transmission electron microscope* (TEM) directs a beam of electrons through a specimen. Electrons are emitted by a tungsten cathode when it is electrically heated. The electric potential of the cathode is kept at 50,000 – 100,000 volts; that of the anode, near the top of the tube, is zero. This drop in voltage causes the electrons to accelerate as they move toward the anode. A condenser lens focuses the electron beam onto the sample; objective and projector lenses focus the electrons that pass through the specimen and project them onto a viewing screen or a piece of photographic film.

In typical electron microscopes, electrons have the properties of a wave with a wavelength of only 0.005 nm.

Recall that the minimum distance *D* at which two objects can be distinguished is proportional to the wavelength λ of the light that illuminates the objects. Thus the limit of resolution for the electron microscope is theoretically 0.005 nm (less than the diameter of a single atom), or 40,000 times better than the resolution of the light microscope and 2 million times better than that of the unaided human eye. However, the effective resolution of the electron microscope in the study of biological systems is considerably less than this ideal. Under optimal conditions, a resolution of 0.10 nm can be obtained with transmission electron microscopes, about 2000 times better than the best resolution of light microscopes.

PREPARATION OF FIXED, STAINED SAMPLES FOR TEM

Like the light microscope, the transmission electron microscope is used to view thin sections of a specimen, but the fixed sections must be much thinner for electron microscopy (only 50 – 100 nm, about 0.2 percent of the thickness of a single cell). Clearly, only a small portion of a cell can be observed in any one section. Generation of the image depends on differential scattering of the incident electrons by molecules in the preparation. Without staining, the beam of electrons passes through a cell or tissue sample uniformly, so the entire sample appears uniformly bright with little differentiation of components. Staining techniques are therefore used to reveal the location and distribution of specific materials.

Heavy metals, such as gold or osmium, appear dark on a micrograph because they scatter (diffract) most of the incident electrons; scattered electrons are not focused by the electromagnetic lenses and do not form the image. Osmium tetrox-ide preferentially stains certain cellular components, such as membranes, which appear black in micrographs. Specific proteins can be detected in thin sections by use of electron-dense gold particles coated with protein A, a bacterial protein that binds antibody molecules nonspecifically.

Electron microscopy also is used to obtain information about the shapes of purified viruses, fibers, enzymes, and other subcellular particles. In one technique, called *metal shadowing*, a thin layer of evaporated metal, such as platinum, is laid at an angle on a biological sample. An acid bath dissolves the biological material, leaving a metal replica of its surface, which can then be examined in the transmission electron microscope. Variations in the angle and thickness of the deposited metal allow an image to be formed because some incident electrons will be scattered in various directions rather than pass through the preparation. If the metal is deposited mainly on one side of the sample, for instance, the image seems to have "shadows," where the metal appears dark and the shadows appear light.

CRYOELECTRON MICROSCOPY

Standard electron microscopy cannot be used to study live cells because they are generally too vulnerable to the required conditions and preparatory techniques. In particular, the absence of water causes macromolecules to become denatured and nonfunctional. However, the technique of *cryoelectron microscopy* allows examination of hydrated, unfixed, and unstained biological specimens directly in the transmission electron microscope. In this technique, an aqueous suspension of a sample is applied in an extremely thin film to a grid. After it has been frozen in liquid nitrogen and maintained in this state by means of a special mount, it is observed in the electron microscope. The very low temperature ("196 °C) keeps the water from evaporating, even in a vacuum, and the sample can be observed in detail in its native, hydrated state without shadowing or fixing it. By computer-based averaging of images of hundreds of particles, a three-dimensional model almost to atomic resolution can be generated.

SCANNING ELECTRON MICROSCOPY VISUALIZES DETAILS ON THE SURFACES OF CELLS AND PARTICLES

The *scanning electron microscope* allows the investigator to view the surfaces of unsectioned specimens. These cannot be visualized with transmission equipment because the electrons pass through the entire specimen. The sample is fixed, dried, and coated with a thin layer of a heavy metal, such as platinum, by evaporation in a vacuum; in this case, the sample is rotated so that the platinum is deposited uniformly on the surface. An intense electron beam inside the microscope scans rapidly over the sample. Molecules in the specimen are excited and release secondary electrons that are focused onto a scintillation detector; the resulting signal is displayed on a cathode-ray tube. Because the number of secondary electrons produced by any one point on the sample depends on the angle of the electron beam in relation to the surface, the scanning electron micrograph has a threedimensional appearance. The resolving power of scanning electron microscopes, which is limited by the thickness of the metal coating, is only about 10 nm, much less than that of transmission instruments.

PURIFICATION OF CELLS AND THEIR PARTS

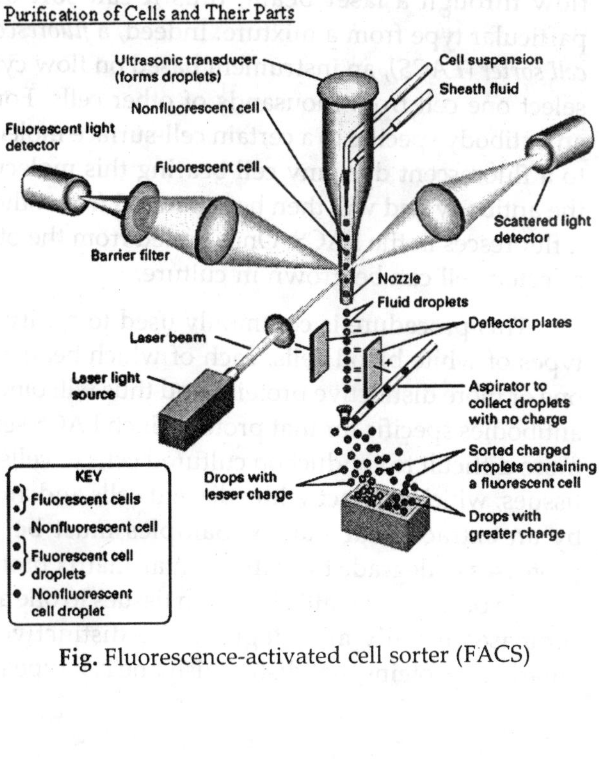

Fig. Fluorescence-activated cell sorter (FACS)

Most animal and plant tissues contain a mixture of cell types. However, an investigator often wishes to study a pure population of one type of cell. In some cases, cells differ in some physical property that allows different cell types to be separated. White blood cells (leukocytes) and red blood cells (erythrocytes), for instance, have very different densities because erythrocytes have no nucleus; thus these cells can be separated on the basis of density. Since most cell types cannot be differentiated so easily, other cell-separation techniques have had to be developed. Similarly, it is essential to isolate quantities of each of the major subcellular organelles to study their structures and metabolic functions in detail.

FLOW CYTOMETRY SEPARATES DIFFERENT CELL TYPES

A *flow cytometer* can identify different cells by measuring the light they scatter, or the fluorescence they emit, as they flow through a laser beam; thus it can sort out cells of a particular type from a mixture. Indeed, a *fluorescence-activated cell sorter (FACS),* an instrument based on flow cytometry, can select one cell from thousands of other cells. For example, if an antibody specific to a certain cell-surface molecule is linked to a fluorescent dye, any cell bearing this molecule will bind the antibody and will then be separated from other cells when it fluoresces in the FACS. Once sorted from the other cells, the selected cell can be grown in culture.

This procedure is commonly used to purify the different types of white blood cells, each of which bears on its surface one or more distinctive proteins and thus will bind monoclonal antibodies specific for that protein. Such FACS separations are more difficult to conduct on cultured cells or cells from animal tissues, which interact with adjacent cells and are surrounded by an extracellular matrix. Samples must be treated with proteases to degrade the extracellular-matrix proteins and cell-surface proteins that attach cells in tissues to one another; these proteases usually also degrade the distinctive cell-surface "marker" proteins that distinguish one cell type from another.

Other uses of flow cytometry include the measurement of a cell's DNA and RNA content and the determination of its general shape and size. The FACS can make simultaneous measurements of the size of a cell (from the amount of scattered light) and the amount of DNA it contains (from the amount of fluorescence from a DNA-binding dye).

DISRUPTION OF CELLS RELEASES THEIR ORGANELLES AND OTHER CONTENTS

The initial step in purifying subcellular structures is to rupture the plasma membrane and the cell wall, if present. First, the cells are suspended in a solution of appropriate pH and salt content, usually isotonic sucrose (0.25 M) or a combination of salts similar in composition to those in the cell's interior. Many cells can then be broken by stirring the cell suspension in a high-speed blender or by exposing it to highfrequency sound *(sonication)*. Plasma membranes can also be sheared by special pressurized tissue homogenizers in which the cells are forced through a very narrow space between the plunger and the vessel wall. Generally, the cell solution is kept at 0 °C to best preserve enzymes and other constituents after their release from the stabilizing forces of the cell.

Because the plasma membrane is highly permeable to water but poorly permeable to the salts and other small molecules (solutes) within cells, *osmotic flow* can be enlisted to help rupture cells. Recall that water flows across a semipermeable membrane, such as the plasma membrane, from a solution of high water (low solute) concentration to one of low water (high solute) concentration until the water concentration on both sides is equal. Consequently, when cells are placed in a hypotonic solution (i.e., one with a lower salt concentration than that of the cell interior), water flows into the cells. This osmotic flow causes the cells to swell and then more easily rupture. Conversely, in a hypertonic solution (i.e., one with a higher salt concentration than that of the cell interior), water flows out of cells, causing them to shrink. When cells are placed in an isotonic solution (i.e., one with a salt concentration equal to that of the cell interior), there is no net movement of water in or out of cells. For this reason, an isotonic solution is best for preserving normal cell structure.

Disrupting the cell produces a mix of suspended cellular components, the *homogenate,* from which the desired organelles can be retrieved. Because rat liver contains an abundance of a single cell type, this tissue has been used in many classic studies of cell organelles. However, the same isolation principles apply to virtually all cells and tissues, and modifications of these cell-fractionation techniques can be used to separate and purify any desired components.

DIFFERENT ORGANELLES CAN BE SEPARATED BY CENTRIFUGATION

We discussed the principles of centrifugation and the uses of centrifugation techniques for separating proteins and nucleic acids. Similar approaches are used for separating and purifying the various organelles, which differ in both size and density.

Most fractionation procedures begin with *differential centrifugation* at increasingly higher speeds also called *differential-velocity centrifugation.* The different sedimentation rates of various cellular components make it possible to separate them partially by centrifugation. Nuclei and viral particles can sometimes be purified completely by such a procedure. After centrifugation at each speed for an appropriate time, the supernatant is poured off and centrifuged at higher speed. Each pelleted fraction can be resuspended and further separated by equilibrium densitygradient centrifugation (discussed next).

Differential centrifugation does not yield totally pure organelle fractions. One method for further purifying fractions is *equilibrium density-gradient centrifugation,* which separates cellular components according to their density. The impure organelle fraction is layered on top of a solution that contains a gradient of a dense nonionic substance, such as sucrose or glycerol. The tube is centrifuged at a high speed (about 40,000 rpm) for several hours, allowing each particle to migrate to an equilibrium position where the density of the surrounding liquid is equal to the density of the particle. In typical preparations from animal cells, the rough endoplasmic reticulum (density = 1.20 g/cm^3) separates well from the Golgi vesicles (density = 1.14 g/cm^3) and from the plasma membrane (density = 1.12 g/cm^3). (The higher density of the rough endoplasmic reticulum is due largely to the ribosomes bound to it.) This method also works well for resolving lysosomes, mitochondria, and peroxisomes in the initial mixed fraction obtained by differential centrifugation.

Since each organelle has unique morphological features, the purity of organelle preparations can be assessed by examination in an electron microscope. Alternatively, organelle-specific marker molecules can be quantified. For example, the protein cytochrome *c* is present only in mitochondria, so the presence of this protein in a fraction of lysosomes would indicate its contamination by mitochondria. Similarly, catalase is present only in peroxisomes; acid phosphatase, only in lysosomes; and ribosomes, only in the rough endoplasmic reticulum or the cytosol.

ORGANELLE-SPECIFIC ANTIBODIES ARE USEFUL IN PREPARING HIGHLY PURIFIED ORGANELLES

Cell fractions often contain more than one type of organelle even after differential and equilibrium density-gradient centrifugation. Such fractions can be further purified by immunological techniques, using monoclonal antibodies for various organelle-specific membrane proteins. One example is the purification of a particular class of cellular vesicles whose outer surface is coated with the protein *clathrin*. An antibody to clathrin, bound to a bacterial carrier, can selectively bind these vesicles in a crude preparation of membranes, and the whole antibody complex can then be isolated by low-speed centrifugation. A recently developed technique uses tiny metallic beads coated with specific antibodies. Organelles that bind to the antibodies, and thus are linked to the metallic beads, are recovered from the preparation by adhesion to a small magnet on the side of the test tube.

All cells contain a dozen or more different types of small membrane-limited vesicles of about the same size (50–100 nm in diameter) and density. Because of their similar size and density, these vesicles are difficult to separate from one another by centrifugation techniques. Immunological techniques are particularly useful for purifying specific classes of such vesicles. Fat and muscle cells, for instance, contain a particular glucose transporter (GLUT4) that is localized to the membrane of a specific kind of vesicle. When insulin is added to the cells, these vesicles fuse with the cell-surface membrane, a process critical to maintaining the appropriate concentration of sugar in the blood. These vesicles can be purified using an antibody that binds to a segment of the GLUT4 protein that faces the cytosol.

BIOMEMBRANES: STRUCTURAL ORGANIZATION AND BASIC FUNCTIONS

Although all biomembranes have the same basic phospholipid bilayer structure and certain common functions, each type of cellular membrane also has certain distinctive activities determined largely by the unique set of proteins associated with that membrane. The two basic categories of membrane proteins were introduced: integral proteins, all or part of which penetrate or span the phospholipid bilayer, and peripheral proteins, which do not interact with the hydrophobic core of the bilayer. In this section, we first discuss the basic principles that govern the organization of phospholipids and integral proteins in all biological membranes and then outline the functions of the plasma membrane in prokaryotes and eukaryotes.

PHOSPHOLIPIDS ARE THE MAIN LIPID CONSTITUENTS OF MOST BIOMEMBRANES

The most abundant lipid components in most membranes are phospholipids, which are amphipathic molecules (i.e., they have a hydrophilic and a hydrophobic part). In *phosphoglycerides,* a principal class of phospholipids, fatty acyl side chains are esterified to two of the three hydroxyl groups in glycerol, and the third hydroxyl group is esterified to phosphate. The phosphate group is also esterified to a hydroxyl group on another hydrophilic compound, such as choline in phosphatidylcholine. Instead of choline, alcohols such as ethanolamine, serine, and the sugar derivative inositol are linked to the phosphate in other phosphoglycerides. The negative charge on the phosphate as well as the charged groups or hydroxyl groups on the alcohol esterified to it interact strongly with water. Both of the fatty acyl side chains in a phosphoglyceride may be saturated or unsaturated, or one chain may be saturated and the other unsaturated.

Sphingomyelin, a phospholipid that lacks a glycerol backbone, is found mainly in plasma membranes. Instead of a glycerol backbone, it contains *sphingosine,* an amino alcohol with a long unsaturated hydrocarbon chain. In sphingomyelin, the terminal hydroxyl group of sphingosine is esterified to phosphocholine, so its hydrophilic head is similar to that of phosphatidylcholine.

Cholesterol and its derivatives constitute another important class of membrane lipids, the steroids. The basic structure of steroids is the four-ring hydrocarbon. Cholesterol, the major steroidal constituent of animal tissues, has a hydroxyl substituent on one ring. Although cholesterol is almost entirely hydrocarbon in composition, it is amphipathic because its hydroxyl group can interact with water. Cholesterol is especially abundant in the plasma membrane of mammalian cells but is absent from most prokaryotic cells. As much as 30 to 50 percent of the lipids in plant plasma membranes consists of cholesterol and certain steroids unique to plants.

Carbohydrates are found in many membranes, covalently bound either to proteins as constituents of glycoproteins or to lipids as constituents of glycolipids. Bound carbohydrates increase the hydrophilic character of lipids and proteins and help to stabilize the conformations of many membrane proteins. The simplest glycolipid, *glucosylcerebroside,* contains a single glucose unit attached to a ceramide.

Every Cellular Membrane Forms a Closed Compartment and Has a Cytosolic and an Exoplasmic Face

Phospholipids of the composition present in cells spontaneously form symmetric sheetlike phospholipid bilayers, which are two molecules thick. The hydrocarbon side chains in each leaflet form a hydrophobic core that is 3 – 4 nm thick in most biomembranes. The various phospholipids differ in the charge carried by the polar head groups at neutral pH: some phosphoglycerides (e.g., phosphatidylcholine and phosphatidylethanolamine) have no net electric charge; others (e.g., phosphatidylglycerol and phosphatidylserine) have a net negative charge. Nonetheless, the polar head groups in all phospholipids can pack together into the characteristic bilayer structure. Sphingomyelins are similar in shape to phosphoglycerides and can form mixed bilayers with them.

Perhaps the most important lesson gleaned from the study of pure phospholipid bilayer membranes is that they spontaneously seal to form closed structures that separate two aqueous compartments. Were a phospholipid bilayer to form a sheet with ends in which the hydrophobic interior were in contact with water, it would be unstable; thus a spherical structure with no ends is the most stable state of a phospholipid bilayer.

Similarly, all cellular membranes are closed structures, surrounding the cell itself or individual compartments. Cellular membranes thus have an *internal face* (the side oriented toward the interior of the compartment) and an *external face* (the side presented to the environment). Because most organelles are surrounded by a single bilayer membrane, it is also useful to speak of the cytosolic face and exoplasmic face of the membrane, the cytosol being the part of the cytoplasm outside of organelles. Thus the exoplasmic face of such organelles faces inward. Similarly, the exoplasmic face of the plasma membrane is directed away from the cytosol, in this case toward the extracellular space, and defines the outer limit of the cell. Some organelles, such as the nucleus, mitochondrion, and chloroplast, are surrounded by two membranes; in these cases, the exoplasmic surface faces the lumen, or space, between the two membranes.

SEVERAL TYPES OF EVIDENCE POINT TO THE UNIVERSALITY OF THE PHOSPHOLIPID BILAYER

A typical cell contains myriad types of membranes, each in turn bearing unique properties bestowed by its particular mix of lipids and proteins. When samples of plasma, nuclear, and mitochondrial membranes are prepared, using the cell-fractionation techniques we have already described, these preparations are often contaminated with the membranes of many other organelles. (The plasma membrane of human erythrocytes, however, can be isolated in near purity because these cells contain no internal membranes.) And all cellular membranes, regardless of their source, possess enormously varied protein-to-lipid ratios. The inner mitochondrial membrane, for example, is 76 percent protein; the myelin membrane, only 18 percent. The high phospholipid content of myelin allows it to electrically insulate the nerve cell from its environment.

Given the variable composition of cellular membranes, how certain are we that the phospholipid bilayer structure is common to all biomembranes? One piece of evidence is that many of the physical properties of pure phospholipid bilayers are similar to those of natural cellular membranes. Another is that either a single species of phospholipid, or a mixture of phospholipids with a composition approximating that found in natural membranes, spontaneously forms either planar bilayers or liposomes when dispersed in aqueous solutions.

Perhaps the best evidence for the bilayer structure comes from low-angle x-ray diffraction analysis of the multimembrane myelin sheath, which is elaborated by Schwann cells and covers and insulates many mammalian nerve cells. The myelin sheath, which is a series of stacked membranes, is the major membrane component of such nerves and can be separated from other cellular membranes in a pure state, permitting direct physical and chemical analyses. X-ray diffraction analysis of these stacked plasma membranes has revealed a regular variation in density that is consistent with a bilayer organization of each membrane unit. In this organization, protein is located mainly on either side of the membrane, which has hydrophilic external faces and a central region of almost pure low-density hydrocarbon. Although some polypeptide segments pass through the lipid bilayer, these make up less than 10 percent of the inner mass of the membrane and are not detected in this type of analysis.

Electron microscopy of thin membrane sections stained with osmium tetroxide, which binds strongly to the polar head groups of phospholipids, provides the most direct evidence for the universality of the bilayer structure. A cross section of all single membranes stained with osmium tetroxide looks like a railroad track: two thin dark lines (the stain – head group complexes) with a uniform light space of about 2 nm (the hydrophobic tails) between them. Although some osmium tetroxide may bind to double bonds in the hydrophobic fatty acyl chains, most of it binds to the polar head groups.

ALL INTEGRAL PROTEINS AND GLYCOLIPIDS BIND ASYMMETRICALLY TO THE LIPID BILAYER

The spaces inside and outside the closed compartments formed by cellular membranes usually have very different compositions. Such *asymmetry* is an essential aspect of the structure and function of biological membranes, and is reflected in the asymmetric structure of all integral membrane proteins. That is, each type of integral membrane protein has a single, specific orientation with respect to the cytosolic and exoplasmic faces of a cellular membrane, and all molecules of any particular integral membrane protein share this orientation. This absolute asymmetry in protein orientation confers different properties on the two membrane faces. Proteins have never been observed to flip-flop across a membrane; such movement, involving a transient movement of hydrophilic amino acid and sugar residues through the hydrophobic interior of the membrane, would be energetically unfavorable. Accordingly, the asymmetry of a membrane protein, which is established during its biosynthesis and insertion into a membrane, is maintained throughout the protein's lifetime.

Both glycoproteins and glycolipids are especially abundant in the plasma membrane of eukaryotic cells but are absent from the inner mitochondrial membrane, the chloroplast lamellae, and several other intracellular membranes. Almost invariably, attached carbohydrates are localized to the exoplasmic membrane face. Glycolipids are always found in the exoplasmic leaflet of membranes and are situated mainly, but not exclusively, on the surface membrane of cells. As with glycoproteins, their polar carbohydrate chains face outward toward the environment and away from the cell.

THE PHOSPHOLIPID COMPOSITION DIFFERS IN TWO MEMBRANE LEAFLETS

Most kinds of phospholipid, as well as cholesterol, are generally present in both membrane leaflets, although they are often more abundant in one or the other. For instance, in plasma membranes from human erythrocytes and certain canine kidney cells grown in culture, almost all the sphingomyelin and phosphatidylcholine, both of which have a positively charged head group are found in the exoplasmic leaflet. In contrast, lipids with neutral or negative polar head groups (e.g., phosphatidylethanolamine, phosphatidylserine, and phosphatidylinositol) are preferentially located in the cytosolic leaflet. Phosphorylated forms of phosphatidylinositol are cleaved as a result of cell stimulation by certain hormones, generating in the cytosol soluble forms of the "head groups" that affect many aspects of cellular metabolism.

The relative abundance of a particular phospholipid in the two leaflets of a plasma membrane can be determined based on its susceptibility to hydrolysis by *phospholipases*, enzymes that cleave the phosphoester bonds that connect the phospholipid head groups. Phospholipids in the cytosolic leaflet are resistant to hydrolysis by phospholipases added to the external medium, because the enzymes cannot penetrate to the cytosolic face of the plasma membrane. It is not clear how these differences in lipid composition of the two leaflets arise. One possibility is that certain lipids bind to specific protein domains that occur preferentially in one membrane leaflet.

Most Lipids and Integral Proteins Are Laterally Mobile in Biomembranes

In both pure phospholipid bilayers and natural membranes, thermal motion permits phospholipid and glycolipid molecules to rotate freely around their long axes and to diffuse laterally within the membrane leaflet. Because such movements are lateral or rotational, the fatty acyl chains remain in the hydrophobic interior of the membrane. In both natural and artificial membranes, a typical lipid molecule exchanges places with its neighbors in a leaflet about 10^7 times per second and diffuses several micrometers per second at 37 °C. At this rate, a lipid could diffuse the length of a typical bacterial cell (≈1 μm) in only 1 second and the length of an animal cell in about 20 seconds.

In pure phospholipid bilayers, phospholipids do not migrate, or flip-flop, from one leaflet of the membrane to the other. In some natural membranes, however, they occasionally do so, catalyzed by certain membrane proteins called *flippases*). Energetically, such movements are extremely unfavorable, because the polar head of a phospholipid must be transported through the hydrophobic interior of the membrane.

Various experiments have shown that many integral membrane proteins, like phospholipids, float quite freely within the plane of a natural membrane. In one such study two different cells (e.g., mouse and human fibroblasts) are fused and the movement of their distinct surface proteins is then monitored at various times after incubation at 37 °C. Such experiments suggest that many integral proteins are free to diffuse in a sea of lipid in the two-dimensional space of the membrane. According to this concept, known as the *fluid mosaic model,* the membrane is viewed as a two-dimensional mosaic of laterally mobile phospholipid and protein molecules. As discussed some integral membrane proteins consist of two or more noncovalently linked subunits; such multimeric membrane proteins float as a unit in the lipid.

The lateral movements of surface proteins and lipids can be quantified by a technique called *fluorescence recovery after photobleaching* (FRAP). With this method the rate at which surface protein or lipid molecules move — the diffusion coefficient — can be determined, as well as the proportion of the molecules that are laterally mobile. FRAP studies with fluorescent-labeled phospholipids have shown that in fibroblast plasma membranes, all the phospholipids are freely mobile over distances of about 0.5 μm, but most cannot diffuse over much longer distances. These findings suggest that protein-rich regions of the plasma membrane, about 1 μm in diameter, separate lipid-rich regions containing the bulk of the membrane phospholipid. Phospholipids are free to diffuse within such a region but not from one lipid-rich region to an adjacent one. Furthermore, the rate of lateral diffusion of lipids in the plasma membrane is nearly an order of magnitude slower than in pure phospholipid bilayers: diffusion constants of 10^{-8} cm^2/s and 10^{-7} cm^2/s are characteristic of the plasma membrane and a lipid bilayer, respectively. This difference suggests that lipids may be tightly but not irreversibly bound to certain integral proteins in some membranes.

Numerous experiments similar to those just discussed have shown that depending on the cell type, 30 – 90 percent of all integral proteins in the plasma membrane are freely mobile. Immobile proteins are permanently attached to the underlying cytoskeleton. The lateral diffusion rate of a mobile protein in an intact membrane is generally 10 – 30 times lower than that of the same protein embedded in synthetic liposomes. These findings suggest that the mobility of integral proteins in intact membranes is restricted by interactions with the rigid submembrane cytoskeleton. Clearly, such interactions would have to be broken and remade as mobile proteins diffuse laterally in the plasma membrane.

FLUIDITY OF MEMBRANES DEPENDS ON TEMPERATURE AND COMPOSITION

One consequence of the packing of the fatty acyl chains within the centre of a phospholipid bilayer is an abrupt change in its physical properties over a very narrow temperature range. For example, when a suspension of liposomes or a planar bilayer composed of a single type of phospholipid is heated, it passes from a highly ordered, gel-like state to a more mobile fluid state. During this *phase transition*, a relatively large amount of heat (thermal energy) is absorbed over a narrow temperature range; the midpoint of this range is the "melting temperature" of the bilayer.

In general, lipids with short or unsaturated fatty acyl chains undergo the phase transition at lower temperatures than do lipids with long or saturated chains. Compared with long chains, short chains have less surface area to form van der Waals interactions with one another. Since the gel state is stabilized by these interactions, short-chain lipids melt at lower temperatures than long-chain lipids. Likewise, the kinks in unsaturated fatty acyl chains result in their forming less stable van der Waals interactions with other lipids than do saturated chains. As a result, unsaturated lipids maintain a more random, fluid state at lower temperatures than lipids with saturated fatty acyl chains.

The hydrophobic interior of natural membranes generally has a low viscosity and a fluidlike, rather than gel-like, consistency. Maintenance of this bilayer fluidity appears to be essential for normal cell growth and reproduction. All cell membranes contain a mixture of different fatty acyl chains and are fluid at the temperature at which the cell is grown. Animal and bacterial cells adapt to a decrease in growth temperature by increasing the proportion of unsaturated to saturated fatty acids in the membrane, which tends to maintain a fluid bilayer at the reduced temperature.

Membrane cholesterol is another major determinant of bilayer fluidity. Cholesterol is too hydrophobic to form a sheet structure on its own, but it is intercalated (inserted) among phospholipids. Its polar hydroxyl group is in contact with the aqueous solution near the polar head groups of the phospholipids; the steroid ring interacts with and tends to immobilize their fatty acyl chains. The net effect of cholesterol on membrane fluidity varies, depending on the lipid composition. Cholesterol restricts the random movement of the polar heads of the fatty acyl chains, which are closest to the outer surfaces of the leaflets, but it separates and disperses their tails, causing the inner regions of the bilayer to become slightly more fluid. At the high concentrations found in eukaryotic plasma membranes, cholesterol tends to make the membrane less fluid at growth temperatures near 37 °C. Below the temperature that causes a phase transition, cholesterol keeps the membrane in a fluid state by preventing the hydrocarbon fatty acyl chains of the membrane lipids from binding to one another, thereby offsetting the drastic reduction in fluidity that would otherwise occur at low temperatures.

MEMBRANE LEAFLETS CAN BE SEPARATED AND EACH FACE VIEWED INDIVIDUALLY

When a frozen tissue specimen is fractured by a sharp blow, the fracture line frequently runs through the hydrophobic interior of cell membranes, separating the two phospholipid leaflets. Integral membrane proteins generally remain associated with one or the other leaflet of membranes subjected to this *freeze-fracturing* technique. The fractured specimen then is placed in a vacuum and the surface ice is removed by sublimation, a technique called *deep etching* or *freeze etching*. After metal shadowing with platinum and carbon, the organic material is removed by acid, leaving a carbon-metal replica of the membrane leaflet.

Electron microscopy of membrane samples prepared by these techniques reveals numerous protuberances, most of which are membrane proteins. In deep-etching studies, the cytoplasmic face of a membrane is customarily called the *P (protoplasmic) face* and the exoplasmic face is the *E face*. It is not unusual for most or all protuberances to be on one of the two surfaces and their mirror images, in the form of pits or holes, to be on the other. This may occur because the integral proteins are bound more tightly to the lipids in one leaflet than to those in the other.

THE PLASMA MEMBRANE HAS MANY COMMON FUNCTIONS IN ALL CELLS

In all cells, the plasma membrane has several essential functions. These include transporting nutrients into and metabolic wastes out of the cell; preventing unwanted materials in the extracellular milieu from entering the cell; preventing loss of needed metabolites and maintaining the proper ionic composition, pH (≈7.2), and osmotic pressure of the cytosol. To carry out these functions, the plasma membrane contains specific *transport proteins* that permit the passage of certain small molecules but not others. Several of these proteins use the energy released by ATP hydrolysis to pump ions and other molecules into or out of the cell against their concentration gradients. Small charged molecules such as ATP and amino acids can diffuse freely within the cytosol but are restricted in their ability to leave or enter it across the plasma membrane.

In addition to these universal functions, the plasma membrane has other crucial roles in multicellular organisms. Few of the cells in multicellular plants and animals exist as isolated entities; rather, groups of cells with related specializations combine to form tissues. Specialized areas of the plasma membrane contain proteins and glycolipids that form specific contacts and junctions between cells to strengthen tissues and to allow the exchange of metabolites between cells. Other proteins in the plasma membrane act as anchoring points for many of the cytoskeletal fibers that permeate the cytosol, imparting shape and strength to cells. Surrounding most animal cells is a mixture of fibrous proteins and polysaccharides collectively called the extracellular matrix. This viscous, water-filled matrix provides a bedding on which most sheets of epithelial cells or small glands lie. Proteins in the plasma membrane anchor cells to many of the matrix components, adding to the strength and rigidity of many tissues. In addition, enzymes bound to the plasma membrane catalyze reactions that would occur with difficulty in an aqueous environment. The plasma membrane of many types of eukaryotic cells also contains receptor proteins that bind specific signaling molecules (e.g., hormones, growth factors, neurotransmitters), leading to various cellular responses. These membrane proteins, which are critical for cell development and functioning, are described in later chapters.

Unlike animal cells, plant cells are surrounded by a cell wall and lack the extracellular matrix found in animal tissues. As a plant cell matures, new layers of wall are laid down just outside the plasma membrane, which is intimately involved in the assembly of cell walls. The walls are built primarily of cellulose, a rodlike polysaccharide formed from β(1→14)-linked glucose monomers. The cellulose molecules aggregate, by hydrogen bonding, into bundles of fibers; other polysaccharides within the wall cross-link the cellulose fibers. In woody plants, a complex water-insoluble polymer of phenol and other aromatic monomers, called *lignin,* imparts strength and rigidity to the cell walls. Other chemicals also are found in the walls of various plant cells; for example, waxes prevent plant tissues and proteins from drying out.

Like the entire cell, each organelle in eukaryotic cells is bounded by a membrane containing a unique set of proteins essential for its proper functioning. In the next section, we discuss the structure and function of the main organelles found in eukaryotic cells.

ORGANELLES OF THE EUKARYOTIC CELL

The various techniques described earlier have led to an appreciation of the highly organized internal structure of eukaryotic cells, marked by the presence of many different organelles. Here we present a brief overview of the major organelles. Unique proteins in the interior and membranes of each type of organelle largely determine its specific functional characteristics. Later chapters will examine the key roles that different organelles and the cytosol play in the functioning of eukaryotic cells.

LYSOSOMES ARE ACIDIC ORGANELLES THAT CONTAIN A BATTERY OF DEGRADATIVE ENZYMES

Lysosomes provide an excellent example of the ability of intracellular membranes to form closed compartments in which the composition of the *lumen* (the aqueous interior of the compartment) differs substantially from that of the surrounding cytosol. Found in animal cells, lysosomes are bounded by a single membrane and are responsible for degrading certain components that have become obsolete for the cell or organism. In some cases, materials taken into a cell by endocytosis or phagocytosis also are degraded in lysosomes. Endocytosis refers to the process by which extracellular materials are taken up by invagination of a segment of the plasma membrane to form a small membrane-bounded vesicle (endosome). In phagocytosis, relatively large particles are enveloped by the plasma membrane and internalized.

Lysosomes contain a group of enzymes that degrade polymers into their monomeric subunits. For example, nucleases degrade RNA and DNA into their mononucleotide building blocks; proteases degrade a variety of proteins and peptides; phosphatases remove phosphate groups from mononucleotides, phospholipids, and other compounds; still other enzymes degrade complex polysaccharides and lipids into smaller units. *Tay-Sachs disease* is caused by a defect in one enzyme catalyzing a step in the lysosomal breakdown of certain glycolipids called gangliosides, which are abundant in nerve cells—with devastating consequences. The symptoms of this inherited disease usually are evident before the age of 1. Affected children commonly become demented and blind by age 2, and die before their third birthday. Nerve cells from such children are greatly enlarged with swollen lipid-filled lysosomes.

All the lysosomal enzymes work most efficiently at acid pH values and collectively are termed *acid hydrolases*. A hydrogen ion pump and a Cl^- channel protein in the lysosomal membrane maintain the pH of the interior at ≈4.8. The pump hydrolyzes ATP and uses the released free energy to pump H^+ ions from the cytosol into the lumen of the lysosome; the Cl^- channel allows Xλ ions to enter. Together they transport HCl. The acid pH helps to denature proteins, making them accessible to the action of the lysosomal hydrolases, which themselves are resistant to acid denaturation. Lysosomal enzymes are poorly active at the neutral pH values of cells and most extracellular fluids. Thus if a lysosome releases its contents into the cytosol, where the pH is between 7.0 and 7.3, little degradation of cytosolic components takes place.

Lysosomes vary in size and shape, and several hundred may be present in a typical animal cell. In effect, they function as sites where various materials to be degraded collect. *Primary lysosomes* are roughly spherical and do not contain obvious particulate or membrane debris. *Secondary lysosomes,* which are larger and irregularly shaped, appear to result from the fusion of primary lysosomes with other membrane organelles; they contain particles or membranes in the process of being digested. The process by which an aged organelle is degraded in a lysosome is called *autophagy* ("eating oneself").

PLANT VACUOLES STORE SMALL MOLECULES AND ENABLE THE CELL TO ELONGATE RAPIDLY

Most plant cells contain at least one membrane-limited internal vacuole. The number and size of vacuoles depend on both the type of cell and its stage of development; a single vacuole may occupy as much as 80 percent of a mature plant cell. Plant cells store water, ions, and nutrients such as sucrose and amino acids within these vacuoles. We will see how such materials are accumulated in vacuoles. Vacuoles also act as receptacles for waste products and excess salts taken up by the plant and may function similarly to lysosomes in animal cells. Like lysosomes, vacuoles have an acidic pH, maintained by a proton pump and a Cl^- channel protein in the vacuole membrane, and contain a battery of degradative **enzymes.**

Similar storage vacuoles are found in green algae and many microorganisms such as yeast.

Like most cellular membranes, the vacuolar membrane is permeable to water but is poorly permeable to the small molecules stored within it. Because the solute concentration is much higher in the vacuole lumen than in the cytosol or extracellular fluids, water tends to move by osmotic flow into vacuoles, just as it moves into cells placed in a hypotonic medium. This influx of water causes both the vacuole to expand and water to move into the cell from the wall, creating hydrostatic pressure, or *turgor,* inside the cell. This pressure is balanced by the mechanical resistance of the cellulose-containing cell wall that surrounds plant cells. Most plant cells have a turgor of 5 – 20 atmospheres (atm); their cell walls must be strong enough to react to this pressure in a controlled way. Unlike animal cells, plant cells can elongate extremely rapidly — at rates of 20 – 75 μm/h. This elongation, which usually accompanies plant growth, occurs when a segment of the somewhat elastic cell wall stretches under the pressure created by water taken into the vacuole.

PEROXISOMES DEGRADE FATTY ACIDS AND TOXIC COMPOUNDS

All animal cells (except erythrocytes) and many plant cells contain peroxisomes, a class of small organelles (≈0.2 – 1 μm in diameter) bounded by a single membrane. (*Glyoxisomes* are similar organelles found in plant seeds that oxidize stored lipids as a source of carbon and energy for growth. They contain many of the same types of enzymes as peroxisomes as well as additional ones used to convert fatty acids to glucose precursors.) Peroxisomes contain several *oxidases* — enzymes that use molecular oxygen to oxidize organic substances, in the process forming hydrogen peroxide (H_2O_2), a corrosive substance. Peroxisomes also contain copious amounts of the enzyme *catalase,* which degrades hydrogen peroxide to yield water and oxygen:

$$2H_2O_2 \xrightarrow{catalase} 2H_2O + O_2$$

In contrast to oxidation of fatty acids in mitochondria, which produces CO_2 and is coupled to generation of ATP, peroxisomal oxidation of fatty acids yields acetyl groups and is not linked to ATP formation. The energy released during peroxisomal oxidation is converted to heat, and the acetyl groups are transported into the cytosol, where they are used in the synthesis of cholesterol and other metabolites. In most eukaryotic cells, the peroxisome is the principal organelle in which fatty acids are oxidized, thereby generating precursors for important biosynthetic pathways. Particularly in liver and kidney cells, various toxic molecules that enter the bloodstream also are degraded in peroxisomes, producing harmless products.

In the human genetic disease *X-linked adrenoleukodystrophy* (ADL), peroxisomal oxidation of very long chain fatty acids is defective. The *ADL* gene encodes the peroxisomal membrane protein that transports into peroxisomes an enzyme required for oxidation of these fatty acids. Individuals with the severe form of ADL are unaffected until mid-childhood, when severe neurological disorders appear, followed by death within a few years.

Mitochondria Are the Principal Sites of ATP Production in Aerobic Cells

Most eukaryotic cells contain many mitochondria, which occupy up to 25 percent of the volume of the cytoplasm. These complex organelles, the main sites of ATP production during aerobic metabolism, are among the largest organelles, generally exceeded in size only by the nucleus, vacuoles, and chloroplasts.

Mitochondria contain two very different membranes, an outer one and an inner one, separated by the intermembrane space. The outer membrane, composed of about half lipid and half protein, contains proteins that render the membrane permeable to molecules having molecular weights as high as 10,000. In this respect, the outer membrane is similar to the outer membrane of gram-negative bacteria. The inner membrane, which is much less permeable, is about 20 percent lipid and 80 percent protein — a higher proportion of protein than occurs in other cellular membranes. The surface area of the inner membrane is greatly increased by a large number of infoldings, or *cristae,* that protrude into the *matrix,* or central space.

In nonphotosynthetic cells, the principal fuels for ATP synthesis are fatty acids and glucose. The complete aerobic degradation of glucose to CO_2 and H_2O is coupled to synthesis of as many as 36 molecules of ATP. In eukaryotic cells, the initial stages of glucose degradation occur in the cytosol, where two ATP molecules per glucose molecule are generated. The terminal stages, including those involving phosphorylation coupled to final oxidation by oxygen, are carried out by enzymes in the mitochondrial matrix and cristae. As many as 34 ATP molecules per glucose molecule are generated in mitochondria, although this value can vary because much of the energy released in mitochondrial oxidation can be used for other purposes (e.g., heat generation and the transport of molecules into or out of the mitochondrion), making less energy available for ATP synthesis. Similarly, virtually all the ATP formed during the oxidation of fatty acids to CO_2 is generated in the mitochondrion. Thus the mitochondrion can be regarded as the "power plant" of the cell.

CHLOROPLASTS, THE SITES OF PHOTOSYNTHESIS, CONTAIN THREE MEMBRANE-LIMITED COMPARTMENTS

Except for vacuoles, chloroplasts are the largest and most characteristic organelles in the cells of plants and green algae. They can be as long as 10 μm and are typically 0.5 – 2 μm thick, but they vary in size and shape in different cells, especially among the algae. Like the mitochondrion, the chloroplast is surrounded by an outer and an inner membrane. Chloroplasts also contain an extensive internal system of interconnected membrane-limited sacs called thylakoids, which are flattened to form disks; these often are grouped in stacks called *grana* and embedded in a matrix, the *stroma*. The thylakoid membranes contain green pigments (chlorophylls) and other pigments and enzymes that absorb light and generate ATP during photosynthesis. Part of this ATP is used by enzymes located in the stroma to convert CO_2 into three-carbon intermediates; these are then exported to the cytosol and converted to sugars.

Perhaps surprisingly, the molecular mechanisms by which ATP is formed in mitochondria and chloroplasts are very similar. Chloroplasts and mitochondria share other features: Both often migrate from place to place within cells and also contain their own DNA, which encodes some of the key organellar proteins. The proteins encoded by mitochondrial or chloroplast DNA are synthesized on ribosomes within the organelles. However, most of the proteins in each organelle are encoded in nuclear DNA and are synthesized in the cytosol; these proteins then are incorporated into the organelles.

THE ENDOPLASMIC RETICULUM IS A NETWORK OF INTERCONNECTED INTERNAL MEMBRANES

Generally, the largest membrane in a eukaryotic cell encloses the endoplasmic reticulum (ER) — a compartment comprising a network of interconnected, closed, membrane-bounded vesicles. The endoplasmic reticulum has a number of functions in the cell but is particularly important in the synthesis of many membrane lipids and proteins. The *smooth endoplasmic reticulum* is smooth because it lacks ribosomes; regions of the *rough endoplasmic reticulum* are studded with ribosomes.

THE SMOOTH ENDOPLASMIC RETICULUM

The synthesis of fatty acids and phospholipids occurs in the smooth ER. Although many cells have very little smooth ER, this organelle is abundant in hepatocytes. Enzymes in the smooth ER of the liver modify or detoxify hydrophobic chemicals such as pesticides and carcinogens by chemically converting them into more water-soluble, conjugated products that can be secreted from the body. High doses of such compounds result in a large proliferation of the smooth ER in liver cells.

THE ROUGH ENDOPLASMIC RETICULUM

Ribosomes bound to the rough ER synthesize certain membrane and organelle proteins and virtually all proteins to be secreted from the cell. The ribosomes that fabricate secretory proteins are bound to the rough ER by the nascent polypeptide chain of the protein. As the growing secretory polypeptide emerges from the ribosome, it passes through the rough ER membrane, with the help of specific proteins in the membrane. The newly made secretory proteins accumulate in the lumen (inner cavity) of the rough ER before being transported to their next destination.

All eukaryotic cells contain a discernible amount of rough ER because it is needed for the synthesis of plasma-membrane proteins and proteins of the extracellular matrix. Rough ER is particularly abundant in cells that are specialized to produce secreted proteins. For example, plasma cells produce antibodies, which circulate in the bloodstream, and pancreatic acinar cells synthesize digestive enzymes, which are transported to the intestine via a series of progressively larger ducts. In both types of cells, a large part of the cytosol is filled with rough ER.

GOLGI VESICLES PROCESS AND SORT SECRETORY AND MEMBRANE PROTEINS

Several minutes after proteins are synthesized in the rough ER, most of them leave the organelle within small membrane-bounded transport vesicles. These vesicles, which bud off from regions of the rough ER not coated with ribosomes, carry the proteins to the luminal cavity of another membrane-limited organelle, the Golgi complex, a series of flattened sacs located near the nucleus in many cells.

Three-dimensional reconstructions from serial sections of a Golgi complex reveal a series of flattened membrane vesicles or sacs, surrounded by a number of more or less spherical membrane vesicles. The stack of flattened Golgi sacs has three defined regions — the *cis,* the *medial,* and the *trans.* Transfer vesicles from the rough ER fuse with the cis region of the Golgi complex, where they deposit their proteins. These proteins then progress from the cis to the medial to the trans region. Within each region are different enzymes that modify secretory and membrane proteins differently, depending on their structures and their final destinations.

After secretory proteins are modified in the Golgi sacs, they are transported out of the complex by a second set of transport vesicles, which seem to bud off the trans side of the Golgi complex. Some of these transport vesicles, termed *coated vesicles,* are surrounded by an outer protein cage composed primarily of the fibrous protein clathrin. Some vesicles contain membrane proteins destined for the plasma membrane; others, proteins for lysosomes or for other organelles. How intracellular transport vesicles "know" which membranes to fuse with and where to deliver their contents is also discussed.

THE DOUBLE-MEMBRANED NUCLEUS CONTAINS THE NUCLEOLUS AND A FIBROUS MATRIX

The nucleus, the largest organelle in eukaryotic cells, is surrounded by two membranes, each one a phospholipid bilayer containing many different types of proteins. The inner nuclear membrane defines the nucleus itself. In many cells, the outer nuclear membrane is continuous with the rough endoplasmic reticulum, and the space between the inner and outer nuclear membranes is continuous with the lumen of the rough endoplasmic reticulum.

The two nuclear membranes appear to fuse at the nuclear pores. The distribution of nuclear pores is particularly vivid when the nucleus is viewed by the freeze-fracture technique described earlier. Constructed of a specific set of membrane proteins, these ringlike pores function as channels that regulate the movement of material between the nucleus and the cytosol.

In a growing or differentiating cell, the nucleus is metabolically active, producing DNA and RNA. The latter is exported through nuclear pores to the cytoplasm for use in protein synthesis. In mature erythrocytes from nonmammalian vertebrates and other types of "resting" cells, the nucleus is inactive or dormant and minimal synthesis of DNA and RNA takes place.

In a nucleus that is not dividing, the chromosomes are dispersed and not thick enough to be observed in the light microscope. Only during cell division are chromosomes visible by light microscopy. However, a suborganelle of the nucleus, the nucleolus, is easily recognized under the light microscope. Most of the cell's ribosomal RNA is synthesized in the nucleolus; some ribosomal proteins are added to ribosomal RNAs within the nucleolus as well. The finished or partly finished ribosomal subunit passes through a nuclear pore into the cytosol.

In the electron microscope, the nonnucleolar regions of the nucleus, called the *nucleoplasm,* can be seen to have areas of high DNA concentration, often closely associated with the nuclear membrane. Fibrous proteins called lamins form a two-dimensional network along the inner surface of the inner membrane, giving it shape and apparently binding DNA to it. The breakdown of this network occurs early in cell division.

THE CYTOSOL CONTAINS MANY PARTICLES AND CYTOSKELETAL FIBERS

Because sections for standard electron microscopy must be thinner than 0.1 µm, fibers in the cytosol, which may be several microns in length, appear as long elements only in sections that by chance happen to be in the plane of the fiber bundles. Serial sectioning of a tissue sample can compensate for these shortcomings by tracing a fiber from one image into the next to reconstruct its threedimensional architecture. Because 200 sections are needed to examine a cell 20 µm thick, serial sectioning is a tedious technique. However, sections up to 1 µm thick can be viewed in high-voltage electron microscopes, considerably reducing the number of sections needed to reconstruct three-dimensional images.

Transmission electron micrographs of cytosolic fibers obtained from unsectioned cells reveal an extensive network of microfilaments, microtubules, and intermediate filaments.

These cytoskeletal fibers crisscross one another in complex patterns so that different types of fibers contact one another at many points. In cultured cells, actin microfilaments often occur in bundles of long fibers that appear to be connected by small fibrous proteins.

The cytosol of many cells also contains *inclusion bodies*, granules that are not bounded by a membrane. For instance, muscle cells and hepatocytes contain cytosolic granules of glycogen a glucose polymer that functions as a storage form of usable cellular energy. In well-fed animals, glycogen can account for as much as 10 percent of the wet weight of the liver. The cytosol of the specialized fat cells in adipose tissue contains large droplets of almost pure triacylglycerols, a storage form of fatty acids.

In addition, the cytosol is a major site of cellular metabolism and contains a large number of different enzymes. About 20 – 30 percent of the cytosol is protein, and from a quarter to half of the total protein within cells is in the cytosol. Because of the high concentration of cytosolic proteins, organized complexes of proteins can form even if the energy that stabilizes them is weak. Many investigators believe that the cytosol is highly organized, with most proteins either bound to fibers or otherwise localized in specific regions.

Chapter 6

Manipulating Cells and Viruses in Culture

Understanding modern molecular cell biology requires a familiarity with the most commonly used biological materials and knowledge of the latest experimental techniques. In this chapter, we describe the primary techniques used to culture and manipulate certain cells, viruses, and experimental organisms. Biologists choose to study these cells and organisms because they provide experimentally favorable examples of important molecular events or processes such as the control of gene activity and cell replication, formation of organelles, secretion of proteins, and differentiation of cells.

In the first part of the chapter, we discuss the culture of microbial cells, animal cells, and hybrid cells in vitro and give some examples of their experimental uses. Many studies involve cells or organisms with mutations affecting a specific biochemical or developmental pathway. In such studies, an important criterion for using a particular cell type or organism is the ease with which mutants can be isolated and characterized and the mutated genes identified and cloned. Because viruses contain a small number of genes (4 to ≈100 in different types of viruses), viral mutants commonly are used to study the action of a limited set of genes designed to carry out restricted, specific molecular tasks. To provide a basis for understanding such studies, we consider the basic properties of viruses and their uses in various types of research in the latter part of this chapter.

GROWTH OF MICROORGANISMS IN CULTURE

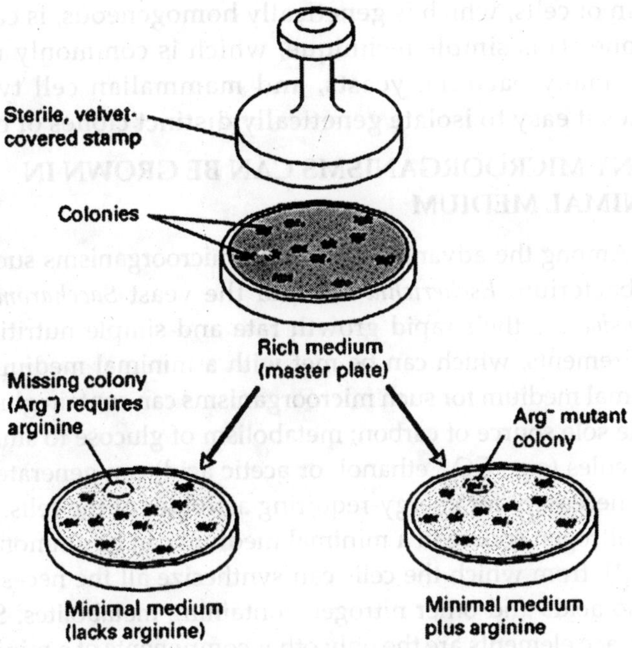

Fig. Replica plating is used to detect random mutations in bacterial and yeast cultures that produce cells differing in a single genetic trait from other cells in the culture.

Cultured cells have several advantages over intact organisms for research on fundamental aspects of cell biology. First, most animal and plant tissues comprise a variety of different types of cells, whereas cells of a specific type can be grown in culture; thus cultured cells are more homogeneous in their properties than in vivo cells in tissues. Second, experimental conditions can be controlled much more rigorously in culture than in an organism. By manipulating the growth conditions, for example, one can measure the effects of specific chemicals and growth factors on a particular cell type in culture. A third reason for preferring cultured cells is that in many cases a single cell can be readily grown into a colony, a process called *cell cloning,* or simply *cloning*. The resulting strain of cells, which is genetically homogeneous, is called a clone. This simple technique, which is commonly used with many bacteria, yeasts, and mammalian cell types, makes it easy to isolate genetically distinct clones of cells.

MANY MICROORGANISMS CAN BE GROWN IN MINIMAL MEDIUM

Among the advantages of using microorganisms such as the bacterium *Escherichia coli* and the yeast *Saccharomyces cerevisiae* are their rapid growth rate and simple nutritional requirements, which can be met with a minimal medium. A minimal medium for such microorganisms can contain glucose as the sole source of carbon; metabolism of glucose to smaller molecules (e.g., CO_2, ethanol, or acetic acid) can generate the ATP necessary for energy-requiring activities of the cells. The sole nitrogen source in a minimal medium can be ammonium (NH_4^+), from which the cells can synthesize all the necessary amino acids and other nitrogen-containing metabolites. Salts and trace elements are the only other components of a minimal medium.

Many prokaryotes (i.e., bacteria) and single-celled eukaryotes such as yeast, both of which grow in nature as single cells, are easily grown in culture dishes — usually on top of *agar,* a semisolid base of plant polysaccharides. The agar is first dissolved in a heated nutrient medium, and the solution is poured into petri dishes; as it cools, the solution solidifies. A dilute suspension of cells then is dispersed on top of the agar; in time each cell grows into a discrete colony. Since the cells in a colony all derive from a single cell, they form a clone and have identical genomes (DNA). All the cells in a clone generally express the same set of genes and contain the same enzymes and other constituents in similar proportions. The division time for *E. coli* and similar microorganisms ranges from 20 minutes to 1 hour. Thus a single *E. coli* cell, which divides approximately every 30 minutes, can grow into a colony containing $10^7 - 10^8$ cells in 12 hours ($2^{24} = 1.7 \times 10^7$).

Because yeasts, unlike most eukaryotic organisms, grow as single cells and can be grown in a simple defined medium similarly to bacteria, they are a popular choice for studies of eukaryotic cell function. The entire life cycle of yeasts can be studied in culture, and colonies can be grown from a single vegetative cell (a growing cell) or from a single spore (a dormant cell). Studies with yeasts have provided valuable information on such subjects as the cellular mechanisms for controlling DNA synthesis, genetic recombination, and the gene products that regulate the cycle of events leading to replication and segregation of chromosomes and ultimately cell division.

MUTANT STRAINS OF BACTERIA AND YEAST CAN BE ISOLATED BY REPLICA PLATING

If a random mutation in a microorganism cultured, or plated, on a minimal agar medium renders cells unable to synthesize a necessary metabolite, and this metabolite is not supplied in the medium (e.g., the amino acid arginine), then the cells will die. But if the cells are grown on a rich medium that supplies the metabolite whose synthesis is prevented by a mutation, then the mutant cell can survive and grow into a colony in which all the cells have the same mutation.

A mutant colony, say, an Arg″ colony, can be detected by *replica plating,* a technique developed by Joshua and Esther Lederberg in the 1940s. The initial plate of multiple colonies growing on rich medium is called the *master plate.* A circular, velvet-covered stamp equal in size to the master plate is pressed onto it, so that the stamp picks up cells from each colony. The stamp is then used to deposit cells onto a second plate containing minimal agar medium without arginine (Arg″ medium) and also onto a third plate containing minimal agar medium with arginine (Arg^+ medium). Cells that can synthesize arginine grow into colonies on both the Arg″ and Arg⁺ plates. However, a clone of cells on the master plate with a mutation in a gene required for arginine synthesis cannot grow on the Arg″ medium, resulting in a space on the Arg″ plate corresponding to the position of this clone on the master plate and Arg⁺ plate. This clone of cells, thus identified as being defective in the synthesis of arginine, then can be isolated from the master plate.

Repeated replica plating of a master plate with paired minimal-medium plates containing or lacking different specific nutrients (e.g., amino acids, nucleic acid bases, vitamins) can identify clones of cells with mutations in the genes required for synthesis of different biological molecules. This approach has been used to isolate thousands of different bacterial and yeast nutritional auxotrophs; these are strains that are unable to grow in the absence of a specific nutrient.

Under standard culture conditions, random mutations occur in *E. coli* genes at a frequency of 1 in 10^6 – 10^8 base pairs per generation. Thus, by the time a single cell has divided to produce a few hundred or thousand cells, some individual genes will differ among cells in the clone. To maintain the genetic homogeneity of a clone, it must be recloned frequently under selective conditions that permit only cells with a particular set of desired characteristics to survive and grow. For example, to maintain an Arg″ clone, isolated as described above, the cells are replated on minimal medium plus arginine; only cells that can synthesize all the metabolites required for growth except arginine will grow on this selective medium.

GROWTH OF ANIMAL CELLS IN CULTURE

Animal cells are more difficult to culture than microorganisms because they require many more nutrients and typically grow only when attached to specially coated surfaces. Despite these difficulties, various types of animal cells, including both undifferentiated and differentiated ones, can be cultured successfully.

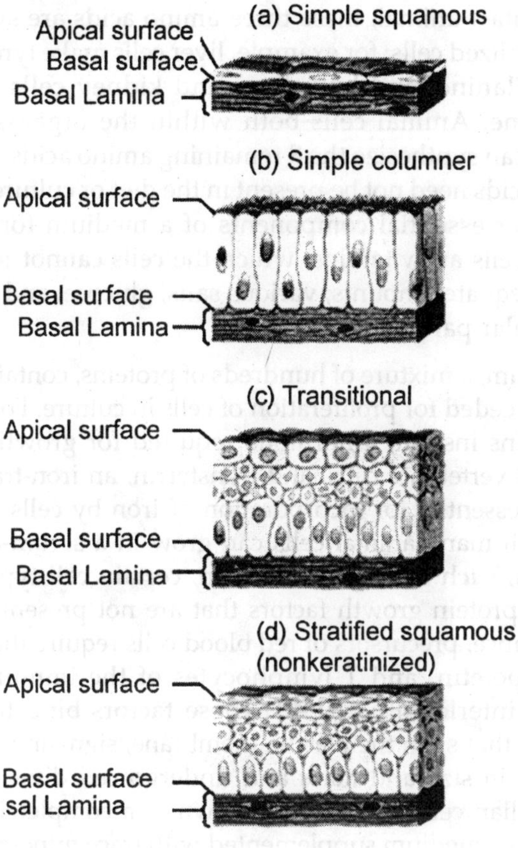

Fig. Principal types of epithelium.

RICH MEDIA ARE REQUIRED FOR CULTURE OF ANIMAL CELLS

Nine amino acids, referred to as the *essential amino acids,* cannot be synthesized by adult vertebrate animals and thus must be obtained from their diet. Animal cells grown in culture also must be supplied with these nine amino acids, namely, histidine, isoleucine, leucine, lysine, methionine, phenylalanine, threonine, tryptophan, and valine. In addition, most cultured cells require cysteine, glutamine, and tyrosine.

In the intact animal, these three amino acids are synthesized by specialized cells; for example, liver cells make tyrosine from phenylalanine, and both liver and kidney cells can make glutamine. Animal cells both within the organism and in culture can synthesize the 8 remaining amino acids; thus these amino acids need not be present in the diet or culture medium. The other essential components of a medium for culturing animal cells are vitamins, which the cells cannot make at all or in adequate amounts; various salts; glucose; and *serum*, the noncellular part of the blood.

Serum, a mixture of hundreds of proteins, contains various factors needed for proliferation of cells in culture. For example, it contains insulin, a hormone required for growth of many cultured vertebrate cells, and transferrin, an iron-transporting protein essential for incorporation of iron by cells in culture. Although many animal cells can grow in a serum-containing medium, such as Eagle's medium, certain cell types require specific protein growth factors that are not present in serum. For instance, precursors of red blood cells require the hormone erythropoietin, and T lymphocytes of the immune system require interleukin 2 (IL-2). These factors bind to receptor proteins that span the plasma membrane, signaling the cells to increase in size and mass and undergo cell division. A few mammalian cell types can be grown in a completely defined, serum-free medium supplemented with trace minerals, specific protein growth factors, and other components.

MOST CULTURED ANIMAL CELLS GROW ONLY ON SPECIAL SOLID SURFACES

Within the tissues of intact animals, most cells tightly contact and interact specifically with other cells via various cellular junctions. The cells also contact the extracellular matrix, a complex network of secreted proteins and carbohydrates that fills the spaces between cells. The matrix, whose constituents are secreted by cells themselves, helps bind the cells in tissues together; it also provides a lattice through which cells can move, particularly during the early stages of animal differentiation.

The extracellular matrices in various animal tissues consist of several common components: fibrous collagen proteins; hyaluronan (or hyaluronic acid), a large mucopolysaccharide; and covalently linked polysaccharides and proteins in the form of proteoglycans (mostly carbohydrate) and glycoproteins (mostly protein). However, the exact composition of the matrix in different tissues varies, reflecting the specialized function of a tissue. In connective tissue, for example, the major protein of the extracellular matrix is a type of collagen that forms insoluble fibers with a very high tensile strength. Fibroblasts, the principal cell type in connective tissue, secrete this type of collagen as well as the other matrix components. Receptor proteins in the plasma membrane of cells bind various matrix elements, imparting strength and rigidity to tissues.

The tendency of animal cells in vivo to interact with one another and with the surrounding extracellular matrix is mimicked in their growth in culture. Unlike bacterial and yeast cells, which can be grown in suspension, most cultured animal cells require a surface to grow on. Many types of cells can adhere to and grow on glass, or on specially treated plastics with negatively charged groups on the surface (e.g., SO_3^{2-}). The cultured cells secrete collagens and other matrix components; these bind to the culture surface and function as a bridge between it and the cells. Cells cultured from single cells on a glass or a plastic dish form visible colonies in 10 – 14 days. Some tumor cells can be grown in suspension, a considerable experimental advantage because equivalent samples are easier to obtain from suspension cultures than from colonies grown in a dish.

PRIMARY CELL CULTURES ARE USEFUL, BUT HAVE A FINITE LIFE SPAN

Normal animal tissues (e.g., skin, kidney, liver) or whole embryos commonly are used to establish *primary cell cultures.* To prepare tissue cells for culture (or to remove adherent cells from a culture dish for biochemical studies), trypsin or another protease is used to destroy the proteins in the junctions that normally interconnect cells. For many years, most cell types were difficult, if not impossible, to culture. But the identification and preparation of various protein growth factors that stimulate the replication of specific cell types, as well as other recent modifications in culture methods, now permit experimenters to grow various types of specialized cells.

Many studies with vertebrate cells, however, still are performed with those few cell types that grow most readily in culture. These are not cells of a defined type; rather, they represent whatever grows when a tissue or an embryo is placed in culture. The cell type that usually predominates in such cultures is called a fibroblast because it secretes the types of proteins associated with fibroblasts in fibrous connective tissue of animals. Cultured fibroblasts have the morphology of tissue fibroblasts, but they retain the ability to differentiate into other cell types; thus they are not as differentiated as tissue fibroblasts.

Some studies are conducted with primary cultures of epithelial cells. In general, external and internal surfaces of tissues and organs are covered by a layer of epithelial cells called an epithelium. These highly differentiated cells are said to be *polarized* because the plasma membrane is organized into at least two discrete regions. For example, the epithelial cells that line the intestine form a simple columnar epithelium. That portion of the plasma membrane facing the lumen of the intestine, the *apical* surface, is specialized for absorption; the rest of the plasma membrane, the *basolateral* surface, mediates transport of nutrients from the cell to the blood and forms junctions with adjacent cells and the underlying extracellular matrix called the basal lamina.

Certain cells cultured from blood, spleen, or bone marrow adhere poorly, if at all, to a culture dish but nonetheless grow well. In the body, such nonadherent cells are held in suspension (in the blood), or they are loosely adherent (in the bone marrow and spleen). Because these cells often come from immature stages in the development of differentiated blood cells, they are very useful for studying normal blood cell differentiation and the abnormal development of leukemias.

When cells are removed from an embryo or an adult animal, most of the adherent ones grow continuously in culture for only a limited time before they spontaneously cease growing. Such a culture eventually dies out after many cell doublings, even if it is provided with fresh supplies of all the known nutrients that cells need to grow, including serum. For instance, when human fetal cells are explanted into cell culture, the majority of cells die within a relatively short time; "fibroblasts," although also destined to die, proliferate for a while and soon become the predominant cell type. They divide about 50 times before they cease growth. Starting with 10^6 cells, 50 doublings can produce $10^6 \times 2^{50}$, or more than 10^{20} cells, which is equivalent to the weight of about 10^5 people. Thus, even though its lifetime is limited, a single culture, if carefully maintained, can be studied through many generations. Such a lineage of cells originating from one initial primary culture is called a cell strain.

TRANSFORMED CELLS CAN GROW INDEFINITELY IN CULTURE

To be able to clone individual cells, modify cell behavior, or select mutants, biologists often want to maintain cell cultures for many more than 100 doublings. This is possible with cells derived from some tumors and with rare cells that arise spontaneously because they have undergone genetic changes that endow them with the ability to grow indefinitely. The genetic changes that allow these cells to grow indefinitely are collectively called *oncogenic* transformation, and the cells are said to be *oncogenically transformed*, or simply *transformed*. A culture of cells with an indefinite life span is considered immortal; such a culture is called a cell line to distinguish it from an impermanent cell *strain*.

The ability of cultured cells to grow indefinitely or their tendency to be transformed varies depending on the animal species from which the cells originate. Normal chicken cells rarely are transformed and die out after only a few doublings; even tumor cells from chickens almost never exhibit immortality. Among human cells, only tumor cells grow indefinitely. The HeLa cell, the first human cell type to be grown in culture, was originally obtained in 1952 from a malignant tumor (carcinoma) of the uterine cervix. This cell line has been invaluable for research on human cells.

In contrast to human and chicken cells, cultures of embryonic adherent cells from rodents routinely give rise to cell lines. When adherent rodent cells are first explanted, they grow well, but after a number of serial replatings they lose growth potential and the culture goes into crisis. During this period most of the cells die, but often a rapidly dividing variant cell arises spontaneously and takes over the culture. A cell line derived from such a variant will grow forever if it is provided with the necessary nutrients. Cells in spontaneously established rodent cell lines and in cell lines derived from tumors often have abnormal chromosomes. In addition, their chromosome number usually is greater than that of the normal cell from which they arose, and it continually expands and contracts in culture. Such cells are said to be *aneuploid* (i.e., have an inappropriate number of chromosomes) and are obviously mutants.

Although most cell lines are undifferentiated, some can carry out many of the functions characteristic of the normal differentiated cells from which they are derived. One example is certain hepatoma cell lines (e.g., HepG2) that synthesize most of the serum proteins made by normal hepatocytes (the major cell type in the liver) from which they are derived. These highly differentiated hepatoma cells are often studied as models of normal hepatocytes. Cultured *myoblasts* (muscle precursor cells) are another example of transformed cells that continue to perform many functions of a specialized, differentiated cell. When grown in culture, transformed myoblasts can be induced to fuse to form myotubes. These resemble differentiated multinucleated muscle cells and synthesize many of, if not all, the specialized proteins associated with contraction. Certain lines of epithelial cells also have been cultured successfully. One such line, Madin-Darby canine kidney (MDCK) cells, forms a continuous sheet of polarized epithelial cells one cell thick that exhibits many of the properties of the normal canine kidney epithelium from which it was derived. This type of preparation has proved valuable as a model for studying the functions of epithelial cells.

FUSION OF CULTURED ANIMAL CELLS CAN YIELD INTERSPECIFIC HYBRIDS USEFUL IN SOMATIC-CELL GENETICS

Cultured animal cells infrequently undergo cell fusion spontaneously. The fusion rate, however, increases greatly in the presence of certain viruses that have a lipoprotein envelope similar to the plasma membrane of animal cells. A mutant viral glycoprotein in the envelope promotes cell fusion the mechanism of this effect is discussed. Cell fusion also is promoted by polyethylene glycol, which causes the plasma membranes of adjacent cells to adhere to each other and to fuse. As most fused animal cells undergo cell division, the nuclei eventually fuse, producing viable cells with a single nucleus that contains chromosomes from both "parents." The fusion of two cells that are genetically different yields a hybrid cell called a heterokaryon.

Because some somatic cells from animals can be cultured from single cells in a well-defined medium, it is possible to select for genetically distinct cultured animal cells, just as is done with bacterial and yeast cells. Moreover, during mitosis the chromosomes in an animal cell are large and highly visible after staining, making it easy to distinguish individual chromosomes. Genetic studies of cultured animal cells are called *somatic-cell genetics* to distinguish them from *classical genetics,* which deals with whole organisms derived from germ cells (sperm and eggs).

Cultured cells from different mammals can be fused to produce interspecific hybrids, which have been widely used in somatic-cell genetics. For instance, hybrids can be prepared from human cells and mutant mouse cells that lack an enzyme required for synthesis of a particular essential metabolite. As the human-mouse hybrid cells grow and divide, they gradually lose human chromosomes in random order, but retain the mouse chromosomes. In a medium that can support growth of both the human cells and mutant mouse cells, the hybrids eventually lose all human chromosomes. However, in a medium lacking the essential metabolite that the mouse cells cannot produce, the one human chromosome that contains the gene encoding the needed enzyme will be retained, because any hybrid cells that lose it following mitosis will die. All other human chromosomes eventually are lost.

By using different mutant mouse cells and media in which they cannot grow, researchers have prepared various panels of hybrid cell lines. Each cell line in a panel contains either a single human chromosome or a small number of human chromosomes, and a full set of mouse chromosomes. Because each chromosome can be identified visually under a light microscope, such hybrid cells provide a means for assigning, or "mapping," individual genes to specific chromosomes. For example, suppose a hybrid cell line is shown microscopically to contain a particular human chromosome. That hybrid cell line can then be tested biochemically for the presence of various human enzymes, exposed to specific antibodies to detect human surface antigens, or subjected to DNA hybridization and cloning techniques to locate particular human DNA sequences. The genes encoding a human protein or containing a human DNA sequence detected in such tests must be located on the particular human chromosome carried by the cell line being tested. Panels of hybrids between normal mouse and mutant hamster cells also have been established; in these hybrid cells, the majority of mouse chromosomes are lost, allowing mouse genes to be mapped to specific mouse chromosomes.

HYBRID CELLS OFTEN ARE SELECTED ON HAT MEDIUM

One metabolic pathway has been particularly useful in cell-fusion experiments. Most animal cells can synthesize the purine and pyrimidine nucleotides de novo from simpler carbon and nitrogen compounds, rather than from already formed purines and pyrimidines. The folic acid antagonists amethopterin and aminopterin interfere with the donation of methyl and formyl groups by tetrahydrofolic acid in the early stages of de novo synthesis of glycine, purine nucleoside monophosphates, and thymidine monophosphate. These drugs are called *antifolates,* since they block reactions involving tetrahydrofolate, an active form of folic acid. Many cells, however, contain enzymes that can synthesize the necessary nucleotides from purine bases and thymidine if they are provided in the medium; these *salvage pathways* bypass the metabolic blocks imposed by antifolates.

A number of mutant cell lines lacking the enzyme needed to catalyze one of the steps in a salvage pathway have been isolated. For example, cell lines lacking thymidine kinase (TK) can be selected because such cells are resistant to the otherwise toxic thymidine analog 5-bromodeoxyuridine. Cells containing TK convert 5-bromodeoxyuridine into 5-bromodeoxyuridine monophosphate. This nucleoside mono- phosphate is then converted into a nucleoside triphosphate by other enzymes and is incorporated by DNA polymerase into DNA, where it exerts its toxic effects. This pathway is blocked in cells with a *TK* mutation that prevents production of functional TK enzyme. Hence, TK″ mutants are resis- tant to the toxic effects of 5-bromodeoxyuridine. Similarly, cells lacking the HGPRT enzyme have been selected because they are resistant to the otherwise toxic guanine analog 6-thioguanine. As we will see next, HGPRT″ cells and TK″ cells are useful partners in cell fusions with one another or with cells that have salvage-pathway enzymes but that are differentiated and cannot grow in culture by themselves.

The medium most often used to select hybrid cells is called *HAT medium,* because it contains hypoxanthine (a purine), aminopterin, and thymidine. Normal cells can grow in HAT medium because even though aminopterin blocks de novo synthesis of purines and TMP, the thymidine in the media is transported into the cell and converted to TMP by TK and the hypoxanthine is transported and converted into usable purines by HGPRT. On the other hand, neither TK″ nor HGPRT″ cells can grow in HAT medium because each lacks an enzyme of the salvage pathway. However, hybrids formed by fusion of these two mutants will carry a normal *TK* gene from the HGPRT^{-} parent and a normal *HGPRT* gene from the TK^{-} parent. The hybrids thus will produce both functional salvage-pathway enzymes and grow on HAT medium. Likewise, hybrids formed by fusion of mutant cells and normal cells can grow in HAT medium.

Hybridomas Are Used to Produce Monoclonal Antibodies

Each normal B lymphocyte in an animal is capable of producing a single type of antibody directed against a specific determinant, or epitope, on an antigen molecule. If an animal is injected with an antigen, B lymphocytes that make antibody recognizing the antigen are stimulated to grow and proliferate. Each antigen-activated B lymphocyte forms a clone of cells in the spleen or lymph nodes, with each cell of the clone producing identical antibody, termed monoclonal antibody. Because most natural antigens contain multiple epitopes, exposure of an animal to an antigen usually stimulates formation of several different B-lymphocyte clones, each producing a different antibody; a mixture of antibodies that recognize different epitopes on the same antigen is said to be *polyclonal*.

For many types of studies involving antibodies, monoclonal antibody is preferable to polyclonal antibody. However, biochemical purification of monoclonal antibody from serum is not feasible, in part because the concentration of any given antibody is quite low. For this reason, researchers looked to culture techniques in order to obtain usable quantities of monoclonal antibody. Because primary cultures of normal B lymphocytes do not grow indefinitely, such cultures have limited usefulness for production of monoclonal antibody. This limitation can be avoided by fusing normal B lymphocytes with oncogenically transformed lymphocytes called *myeloma cells,* which are immortal.

Fusion of a myeloma cell with a normal antibody-producing cell from a rat or mouse spleen yields a hybrid that proliferates into a clone called a hybridoma. Like myeloma cells, hybridoma cells are immortal. Each hybridoma produces the monoclonal antibody encoded by its B-lymphocyte partner. Many different myeloma cell lines from mice and rats have been established; from these, HGPRT″ lines have been selected based on their resistance to 6-thioguanine as described above. If such mutant myeloma cells are fused with normal B lymphocytes, any fused cells that result can grow in HAT medium, but the parental cells cannot. Each selected hybridoma then is tested for production of the desired antibody; any clone producing that antibody then is grown in large cultures, from which a substantial quantity of pure monoclonal antibody can be obtained.

Such pure antibodies are very valuable research reagents. For example, a monoclonal antibody that interacts with protein X can be used to label, and thus locate, protein X in specific cells of an organ or in specific cell fractions. Once identified, even very scarce proteins can be isolated by affinity chromatography in columns to which the monoclonal antibody is bound. Monoclonal antibodies also have become important diagnostic and therapeutic tools in medicine. Monoclonal antibodies that bind to and inactivate toxic proteins (toxins) secreted by bacterial pathogens are used to treat diseases caused by these pathogens. Other monoclonal antibodies are specific for cell-surface proteins expressed by certain types of tumor cells; chemical complexes of such monoclonal antibodies with toxic drugs are being developed for cancer chemotherapy.

VIRUSES: STRUCTURE, FUNCTION, AND USES

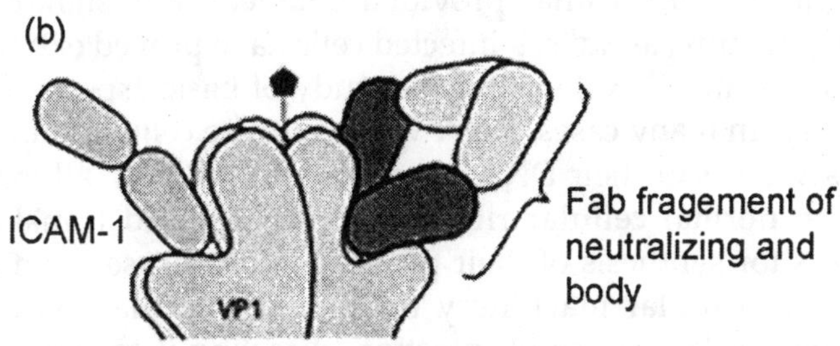

Fig. Structure of picornaviruses.

A virus is a small parasite that cannot reproduce by itself. Once it infects a susceptible cell, however, a virus can direct the cell machinery to produce more viruses. Most viruses have either RNA or DNA as their genetic material. The nucleic acid may be single- or double-stranded. The entire infectious virus particle, called a virion, consists of the nucleic acid and an outer shell of protein. The simplest viruses contain only enough RNA or DNA to encode four proteins. The most complex can encode 100 – 200 proteins.

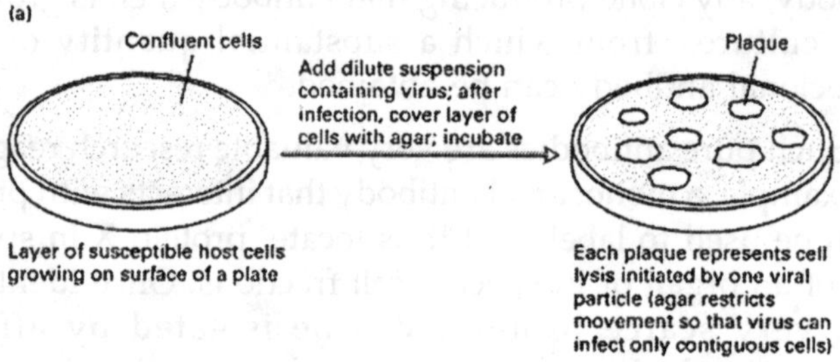

Fig. Plaque assay for determining number of infectious particles in a viral suspension.

The study of plant viruses inspired some of the first experiments in molecular biology. In 1935, Wendell Stanley purified and partly crystallized tobacco mosaic virus (TMV); other plant viruses were crystallized soon thereafter. Pure proteins had been crystallized only a short time before Stanley's work, and it was considered very surprising at the time that a replicating organism could be crystallized.

A wealth of subsequent research with bacterial viruses and animal viruses has provided detailed understanding of viral structure, and virus-infected cells have proved extremely useful as model systems for the study of basic aspects of cell biology. In many cases, DNA viruses utilize cellular enzymes for synthesis of their DNA genomes and mRNAs; all viruses utilize normal cellular ribosomes, tRNAs, and translation factors for synthesis of their proteins. Most viruses commandeer the cellular machinery for macromolecular synthesis during the late phase of infection, directing it to synthesize large amounts of a small number of viral mRNAs and proteins instead of the thousands of normal cellular macromolecules. For instance, animal cells infected by influenza or vesicular stomatitis virus synthesize only one or two types of glycoproteins, which are encoded by viral genes, whereas uninfected cells produce hundreds of glycoproteins. Such virus-infected cells have been used extensively in studies on synthesis of cell-surface glycoproteins. Similarly, much information about the mechanism of DNA replication has come from studies with bacterial cells and animal cells infected with simple DNA viruses, since these viruses depend almost entirely on cellular proteins to replicate their DNA. Viruses also often express proteins that modify host-cell processes so as to maximize viral replication. For example, the roles of certain cellular factors in initiation of protein synthesis were revealed because viral proteins interrupt their action. Finally, when certain genes carried by cancer-causing viruses integrate into chromosomes of a normal animal cell, the normal cell can be converted to a cancer cell.

Since many viruses can infect a large number of different cell types, genetically modified viruses often are used to carry foreign DNA into a cell. This approach provides the basis for a growing list of experimental gene therapy treatments. Because of the extensive use of viruses in cell biology research and their potential as therapeutic agents, we describe the basic aspects of viral structure and function in this section.

VIRAL CAPSIDS ARE REGULAR ARRAYS OF ONE OR A FEW TYPES OF PROTEIN

The nucleic acid of a virion is enclosed within a protein coat, or capsid, composed of multiple copies of one protein or a few different proteins, each of which is encoded by a single viral gene. Because of this structure, a virus is able to encode all the information for making a relatively large capsid in a small number of genes. This efficient use of genetic information is important, since only a limited amount of RNA or DNA, and therefore a limited number of genes, can fit into a virion capsid. A capsid plus the enclosed nucleic acid is called a nucleocapsid.

Nature has found two basic ways of arranging the multiple capsid protein subunits and the viral genome into a nucleocapsid. The simpler structure is a protein helix with the RNA or DNA protected within. Tobacco mosaic virus (TMV) is a classic example of the helical nucleocapsid. In TMV the protein subunits form broken disklike structures, like lock washers, which form the helical shell of a long rodlike virus when stacked together.

The other major structural class of viruses, called *icosahedral* or *quasi-spherical viruses,* is based on the icosahedron, a solid object built of 20 identical faces, each of which is an equilateral triangle. In the simplest type of icosahedral virion each of the 20 triangular faces is constructed of three identical capsid protein subunits, making a total of 60 subunits per capsid. At each of the 12 vertices, five subunits make contact symmetrically. Thus all protein subunits are in *equivalent* contact with one another. Tobacco satellite necrosis virus has such a simple icosahedral structure. However, most quasi-spherical viruses are larger, requiring the assembly of more than three subunits per face of the icosahedron. These proteins form shells whose subunits are in *quasi-equivalent* contact. Here, the proteins at the icosahedral vertices remain arranged in a fivefold symmetry, but additional subunits cover the surfaces between in a pattern of sixfold symmetry.

The atomic structures of a number of icosahedral viruses have been determined by x-ray crystallography. The first three such viruses to be analyzed — tomato bushy stunt virus, poliovirus, and rhinovirus (the common cold virus) — exhibit a remarkably similar design, in terms of the rules of icosahedral symmetry as well as in the details of their surface proteins. In each virus, at atomic resolution, clefts ("canyons") are observed encircling each of the vertices of the icosahedral structure. Interaction of these clefts with cell-surface receptors attaches the virus to a host cell, the first step in viral infection. Neutralizing antibodies specific for a particular virus also interact with these clefts, thereby inhibiting attachment of the virus to the host cell.

In some viruses, the symmetrically arranged nucleocapsid is covered by an external membrane, or *envelope*, which consists mainly of a phospholipid bilayer but also contains one or two types of virus-encoded glycoproteins. The phospholipids in the viral envelope are similar to those in the plasma membrane of an infected host cell. The viral envelope is, in fact, derived by budding from that membrane, but contains mainly viral glycoproteins.

The components of simple viruses such as TMV, which consists of a single RNA molecule and one protein species, undergo self-assembly if they are mixed in solution. More complex viruses containing a dozen or more protein species do not spontaneously assemble in vitro. The multiple components of such viruses assemble within infected cells in stages, first into subviral particles and then into completed virions. The genomes of these complex viruses encode proteins that assist in the assembly of the virion, but the assembly proteins are not themselves components of the completed virion.

MOST VIRAL HOST RANGES ARE NARROW

The fact that the host range — the group of cell types that a virus can infect — is generally restricted serves as a basis for classifying viruses. A virus that infects only bacteria is called a bacteriophage, or simply a phage. Viruses that infect animal or plant cells are referred to generally as *animal viruses* or *plant viruses*. A few viruses can grow in both plants and the insects that feed on them. The highly mobile insects serve as vectors for transferring such viruses between susceptible plant hosts. An example is potato yellow dwarf virus, which can grow in leafhoppers (insects that feed on potato plant leaves) as well as in potato plants. Wide host ranges are characteristic of some strictly animal viruses, such as vesicular stomatitis virus, which grows in insects and in many different types of mammalian cells. Most animal viruses, however, do not cross phyla, and some (e.g., poliovirus) infect only closely related species such as primates. The host-cell range of some animal viruses is further restricted to a limited number of cell types because only these cells have appropriate surface receptors to which the virions can attach.

VIRUSES CAN BE CLONED AND COUNTED IN PLAQUE ASSAYS

The number of infectious viral particles in a sample can be quantified by a plaque assay. This assay is performed by culturing a dilute sample of viral particles on a plate covered with host cells and then counting the number of local lesions, called *plaques,* that develop. A plaque develops on the plate wherever a single virion initially infects a single cell. The virus replicates in this initial host cell and then lyses the cell, releasing many progeny virions that infect the neighboring cells on the plate. After a few such cycles of infection, enough cells are lysed to produce a visible plaque in the layer of remaining uninfected cells.

Since all the progeny virions in a plaque are derived from a single parental virus, they constitute a virus clone. This type of plaque assay is in standard use for bacterial and animal viruses. Plant viruses can be assayed similarly by counting local lesions on plant leaves inoculated with viruses. Analysis of viral mutants, which are commonly isolated by plaque assays, has contributed extensively to current understanding of molecular cellular processes. The plaque assay also is critical in isolating λ bacteriophage clones carrying segments of cellular DNA.

VIRAL GROWTH CYCLES ARE CLASSIFIED AS LYTIC OR LYSOGENIC

The surface of viruses includes many copies of one type of protein that binds, or adsorbs, specifically to multiple copies of a receptor protein on a host cell. This interaction determines the host range of a virus and begins the infection process. Then, in one of various ways, the viral DNA or RNA crosses the plasma membrane into the cytoplasm. The entering genetic material may still be accompanied by inner viral proteins, although in the case of many bacteriophages, all capsid proteins remain outside an infected cell. The genome of most

DNA-containing viruses that infect eukaryotic cells is transported (with some associated proteins) into the cell nucleus, where the cellular DNA is, of course, also found. Once inside the cell, the viral DNA interacts with the host's machinery for transcribing DNA into mRNA. The viral mRNA that is produced then is translated into viral proteins by host-cell ribosomes, tRNA, and translation factors.

Most viral protein products fall into one of three categories: special enzymes needed for viral replication; inhibitory factors that stop host-cell DNA, RNA, and protein synthesis; and structural proteins used in the construction of new virions. These last proteins generally are made in much larger amounts than the other two types. After the synthesis of hundreds to thousands of new virions has been completed, most infected bacterial cells and some infected plant and animal cells rupture, or lyse, releasing all the virions at once. In many plant and animal viral infections, however, no discrete lytic event occurs; rather, the dead host cell releases the virions as it gradually disintegrates.

These events — adsorption, penetration, replication, and release — describe the lytic cycle of viral replication. The outcome is the production of a new round of viral particles and death of the cell. Adsorption and release of enveloped animal viruses are somewhat more complicated processes. In this case, the virions "bud" from the host cell, thereby acquiring their outer phospholipid envelope, which contains mostly viral glycoproteins.

We illustrate the lytic cycle of enveloped viruses with the rabies virus, whose nucleocapsid consists of a single-stranded RNA genome surrounded by multiple copies of nucleocapsid protein. Within the nucleocapsid of rabies virions are viral enzymes for synthesizing viral mRNA and replicating the viral genome. The envelope around the nucleocapsid is a phospholipid bilayer containing multiple copies of a viral transmembrane glycoprotein. This receptor-binding, or "attachment," protein has a large external folded domain on the outside of the viral envelope, an α-helical transmembrane domain that spans the viral envelope, and a short internal domain. The internal domain interacts with the viral matrix protein, which functions as a bridge between the transmembrane glycoprotein and nucleocapsid protein. Assembly of progeny nucleocapsids, and release of progeny virions by budding from the host-cell plasma membrane. Budding virions are clearly visible in electron micrographs.

In some cases, after a bacteriophage DNA molecule enters a bacterial cell, it becomes integrated into the host-cell chromosome, where it remains quiescent and is replicated as part of the cell's DNA from one generation to the next. This association is called lysogeny, and the integrated phage DNA is referred to as a *prophage*. Under certain conditions, the prophage DNA is activated, leading to its excision from the host-cell chromosome and entrance into the lytic cycle. Bacterial viruses of this type are called *temperate phages*. The genomes of a number of animal viruses also can integrate into the host-cell genome. Probably the most important are the retroviruses, described briefly later in this chapter.

A few phages and animal viruses can infect a cell and cause new virion production without killing the cell or becoming integrated.

Four Types of Bacterial Viruses Are Widely Used in Biochemical and Genetic Research

Bacterial viruses have played a crucial role in the development of molecular cell biology. Thousands of different bacteriophages have been isolated; many of these are particularly well suited for studies of specific biochemical or genetic events. Here, we briefly describe four types of bacteriophages, all of which infect *E. coli*, that have been especially useful in molecular biology research.

DNA PHAGES OF THE T SERIES

The T phages of E. coli are large lytic phages that contain a single molecule of double-stranded DNA. This molecule is about 2×10^5 base pairs long in T2, T4, and T6 viruses and about 4×10^4 base pairs long in T1, T3, T5, and T7 viruses. T-phage virions consist of a helical protein "tail" attached to an icosahedral "head" filled with the viral DNA. After the tip of a T-phage tail adsorbs to receptors on the surface of an E. coli cell, the DNA in the head enters the cell through the tail. The phage DNA then directs a programme of events that produces approximately 100 new phage particles in about 20 minutes, at which time the infected cell lyses and releases the new phages. The initial discovery of the role of messenger RNA in protein synthesis was based on studies of E. coli cells infected with bacteriophage T2. By 20 minutes after infection, infected cells synthesize T2 proteins only. The finding that the RNA synthesized at this time had the same base composition as T2 DNA (not E. coli DNA) implied that mRNA copies of T2 DNA were synthesized and used to direct cellular ribosomes to synthesize T2 proteins.

TEMPERATE PHAGES

Bacteriophage λ, which infects E. coli, typifies the temperate phages. This phage has one of the most studied genomes and is used extensively in DNA cloning. On entering an E. coli cell, the double-stranded λ DNA assumes a circular form, which can enter either the lytic cycle (as T phages do) or the lysogenic cycle. In the latter case, proteins expressed from the viral DNA bind a specific sequence on the circular viral DNA to a similar specific sequence on the circular bacterial DNA. The viral proteins then break both circular molecules of DNA and rejoin the broken ends, so that the viral DNA becomes inserted into the host DNA. The carefully controlled action of viral genes maintains λ DNA as part of the host chromosome by repressing the lytic functions of the phage. Under appropriate stimulation, the λ prophage is activated and undergoes lytic replication.

SMALL DNA PHAGES

The genome of some bacteriophages encodes only 10 – 12 proteins, roughly 5 – 10 percent of the number encoded by T phages. These small DNA phages are typified by the Ö×174 and the filamentous M13 phages. These were the first organisms in which the entire DNA sequence of a genome was determined, permitting extensive understanding of the viral life cycle. The viruses in this group are so simple that they do not encode most of the proteins required for replication of their DNA but depend on cellular proteins for this purpose. For this reason, they have been particularly useful in identifying and analyzing the cellular proteins involved in DNA replication.

RNA PHAGES

Some E. coli bacteriophages contain a genome composed of RNA instead of DNA. Because they are easy to grow in large amounts and because their RNA genomes also serve as their mRNA, these phages are a ready source of a pure species of mRNA. In one of the earliest demonstrations that cell-free protein synthesis can be mediated by mRNA, RNA from these phages was shown to direct the synthesis of viral coat protein when added to an extract of E. coli cells containing all the other components needed for protein synthesis. Also, the first long mRNA molecule to be sequenced was the genome of an RNA phage. These viruses, among the smallest known, encode only four proteins: an RNA polymerase for replication of the viral RNA, two capsid proteins, and an enzyme that dissolves the bacterial cell wall and allows release of the intracellular virus particles into the medium.

ANIMAL VIRUSES ARE CLASSIFIED BY GENOME TYPE AND MRNA SYNTHESIS PATHWAY

Animal viruses come in a variety of shapes, sizes, and genetic strategies. In this book, we are concerned with viruses that exhibit at least one of two features: they utilize important cellular pathways to form their molecules, thereby closely mimicking a normal cellular function, or they can integrate their genomes into those of normal cells.

The names of many viruses are based on the names of the diseases they cause or of the animals or plants they infect. Common examples include poliovirus, which causes poliomyelitis; tobacco mosaic virus, which causes a mottling disease of tobacco leaves; and human immunodeficiency virus (HIV), which causes acquired immunodeficiency syndrome

(AIDS). However, many different kinds of viruses often produce the same symptoms or the same apparent disease states; for example, several dozen different viruses can cause the red eyes, runny nose, and sneezing referred to as the common cold. Clearly, any attempt to classify viruses on the basis of the symptoms they produce or their hosts obscures many important differences in their structures and life cycles.

What *are* central to the life cycle of a virus are the types of nucleic acids formed during its replication and the pathway by which mRNA is produced. The relation between the viral mRNA and the nucleic acid of the infectious particle is the basis of a simple means of classifying viruses. In this system, a viral mRNA is designated as a *plus strand* and its complementary sequence, which cannot function as an mRNA, is a *minus strand*. A strand of DNA complementary to a viral mRNA is also a minus strand. Production of a plus strand of mRNA requires that a minus strand of RNA or DNA be used as a template. Using this system, six classes of animal viruses are recognized. Bacteriophages and plant viruses also can be classified in this way, but the system has been used most widely in animal virology because representatives of all six classes have been identified.

DNA VIRUSES (CLASSES I AND II)

The genomes of both class I and class II viruses consist of DNA. Various types of DNA viruses are commonly used in studies on DNA replication, genome structure, mRNA production, and oncogenic cell transformation.

Class I viruses contain a single molecule of double-stranded DNA (dsDNA). In the case of the most common type of class I animal virus, viral DNA enters the cell nucleus, where cellular enzymes transcribe the DNA and process the resulting RNA into viral mRNA. Examples of these viruses include the following:

- *Adenoviruses,* which cause infections in the upper respiratory tract and gastrointestinal tract in many animals

- *SV40* (simian virus 40), a monkey virus that was accidentally discovered in kidney cell cultures from wild monkeys used in the production of poliovirus vaccines
- *Herpesviruses*, which cause various inflammatory skin diseases (e.g., chickenpox) and latent infections that recur after long intervals (e.g., cold sores and shingles)
- *Human papillomaviruses* (HPVs), which cause warts and other insignificant skin lesions and occasionally cause malignant transformation of cervical cells

Some types of HPV are passed through sexual contact. In some infected women, the HPV genome integrates into the chromosome of a cervical epithelial cell. This rare integration event initiates an intensively studied process that can lead to development of cervical carcinoma, one of the most common types of human cancers. Routine Pap smears performed for early detection of cervical carcinoma are done to identify cells in the early stages of the transformation process initiated by HPV integration.

The second type of class I virus, collectively referred to as *poxviruses*, replicates in the host-cell cytoplasm. Typical of class Ib viruses are variola, which causes smallpox, and vaccinia, an attenuated (weakened) poxvirus used in vaccinations to induce immunity to smallpox. These very large, brick-shaped viruses (0.1 × 0.1 × 0.2 μm) carry their own enzymes for synthesizing viral mRNA and DNA in the cytoplasm.

Class II viruses, called *parvoviruses* (from Latin *parvo*, "poor"), are simple viruses that contain one molecule of single-stranded DNA (ssDNA). Some parvoviruses encapsidate (enclose) both plus and minus strands of DNA, but in separate virions; others encapsidate only the minus strand. In both cases, the ssDNA is copied inside the cell into dsDNA, which is then itself copied into mRNA.

RNA VIRUSES (CLASSES III – VI)

All the animal viruses belonging to classes III – VI have RNA genomes. A wide range of animals, from insects to human beings, are infected by viruses in each of these classes. These viruses have been particularly useful in studies on mRNA synthesis and translation (class III); glycoprotein synthesis, membrane formation, and intracellular transport (classes IV and V); and cell transformation and oncogenes (class VI).

Class III viruses contain double-stranded genomic RNA (dsRNA). The minus RNA strand acts as a template for the synthesis of plus strands of mRNA. The virions of all class III viruses known to date have genomes containing 10 – 12 separate double-stranded RNA molecules, each of which encodes one or two polypeptides. Consequently, these viruses are said to have "segmented" genomes. In these viruses, the virion itself contains a complete set of enzymes that can utilize the minus strand of the genomic RNA as a template for synthesis of mRNA in the test tube as well as in the cell cytoplasm after infection. A number of important studies have used class III viruses as a source of pure mRNA.

Class IV viruses contain a single plus strand of genomic RNA, which is identical with the viral mRNA. Since the genomic RNA encodes proteins, it is infectious by itself. During replication of class IV viruses, the genomic RNA is copied into a minus strand, which then acts as a template for synthesis of more plus strands, or mRNA. Two types of class IV viruses are known. In class IVa viruses, typified by poliovirus, viral proteins are first synthesized, from a single mRNA species, as a long polypeptide chain, or *polyprotein,* which is then cleaved to yield the various functional proteins. Class IVb viruses synthesize at least two species of mRNA in a host cell. One of these mRNAs is the same length as the virion's genomic RNA; the other corresponds to the 32 third of the genomic RNA. Both mRNAs are translated into polyproteins. Included in class IVb are a large number of rare insect-borne viruses including Sindbis virus and those causing yellow fever and viral encephalitis in human beings. These viruses once were called *arboviruses* (arthropod-borne viruses), but now are called *togaviruses* (from Latin *toga*, cover) because the virions are surrounded by a lipid envelope.

Class V viruses contain a single negative strand of genomic RNA, whose sequence is complementary to that of the viral mRNA. The genomic RNA in the virion acts as a template for synthesis of mRNA but does not itself encode proteins. Two types of class V viruses can be distinguished. The genome in class Va viruses, which include the viruses causing measles and mumps, is a single molecule of RNA. A virus-specific RNA polymerase present in the virion catalyzes synthesis of several mRNAs, each encoding a single protein, from the genomic template strand. Class Vb viruses, typified by influenza virus, have segmented genomes; each segment acts as a template for the synthesis of a different mRNA species. In most cases, each mRNA produced by a class Vb virus encodes a single protein; however, some mRNAs can be read in two different frames to yield two distinct proteins. As with class Va viruses, a class Vb virion contains a virus-specific polymerase that catalyzes synthesis of the viral mRNA. Thus the genomic RNA (a minus strand) in both types of class V viruses is not infectious in the absence of the virus-specific polymerase. The influenza RNA polymerase initiates synthesis of each mRNA by a unique mechanism. In the host-cell nucleus, the polymerase cuts off 12 – 15 nucleotides from the 52 end of a cellular mRNA or mRNA precursor; this oligonucleotide acts as a "primer" that is elongated by the polymerase to form viral (+) mRNAs, using the genomic (") RNA as a template.

Class VI viruses are enveloped viruses whose genome consists of two identical plus strands of RNA. These viruses are also known as retroviruses because their RNA genome directs the formation of a DNA molecule. The DNA molecule ultimately acts as the template for synthesis of viral mRNA. Initially, a viral enzyme called reverse transcriptase copies the viral RNA genome into a single minus strand of DNA; the same enzyme then catalyzes synthesis of a complementary plus strand. The resulting dsDNA is integrated into the chromosomal DNA of the infected cell. Finally, the integrated proviral DNA is transcribed by the cell's own machinery into (+) RNA, which either is translated into viral proteins or is packaged within virion coat proteins to form progeny virions, which are released by budding from the host-cell membrane. Because most retroviruses do not kill their host cells, infected cells can replicate, producing daughter cells with integrated proviral DNA. These daughter cells continue to transcribe the proviral DNA and bud progeny virions.

Some retroviruses contain cancer-causing genes (called oncogenes). Cells infected by such retroviruses are oncogenically transformed into tumor cells. Studies of oncogenic retroviruses (mostly viruses of birds and mice) have revealed a great deal about the processes that lead to oncogenic transformation. Among the known human retroviruses are human T-cell lymphotrophic virus (HTLV), which causes a form of leukemia, and human immunodeficiency virus (HIV), which causes acquired immune deficiency syndrome (AIDS). Both of these viruses can infect only specific cell types, primarily certain cells of the immune system and, in the case of HIV, some central nervous system neurons and glial cells. Only these cells have cell-surface receptors that interact with viral proteins, accounting for the host-cell specificity of these viruses.

VIRAL VECTORS CAN BE USED TO INTRODUCE SPECIFIC GENES INTO CELLS

Knowledge about mechanisms of viral replication has allowed virologists to modify viruses for various purposes. For instance, the ability of virions to introduce their contents into the cytoplasm and nuclei of infected cells has been adapted for use in DNA cloning and offers possibilities in the treatment of certain diseases. The introduction of new genes into cells by packaging them into virion particles is called *viral gene transduction,* and the virions used for this purpose are called *viral vectors.*

By use of recombinant DNA techniques it is a relatively straightforward process to construct human adenovirus *recombinants* in which potentially therapeutic genes replace the viral genes required for the lytic cycle of infection. Because adenovirus has a very broad host range for different types of human cells, these vectors can introduce the engineered gene into the cells of tissues where they are applied. If the transduced gene encodes the normal form of a protein that is missing or defective in a particular disease, then such *gene therapy* may successfully treat the disease. One type of adenovirus, for example, efficiently infects cells lining the air passages in the lungs, causing a type of common cold. Researchers have replaced some of the disease-causing genes in this adenovirus with the *CFTR* gene, which is defective in individuals with cystic fibrosis. This recombinant adenovirus currently is being used to introduce a normal *CFTR* gene into the airway-lining cells of cystic fibrosis patients. Unfortunately, with most of the adenovirus vectors currently available, the transduced gene usually is expressed only for a limited period of 2 to 3 weeks. This significantly limits their usefulness in gene therapy.

Viral vectors have also been developed from viruses that integrate their genomes into host-cell chromosomes. Such vectors have the advantage that progeny of the initially infected cell also contain and express the transduced gene because it is replicated and segregated to daughter cells along with the rest of the chromosome into which it is integrated. Retroviral vectors, which can efficiently integrate transduced genes at approximately random positions in host-cell chromosomes are now widely used experimentally to generate cultured cells expressing specific, desired proteins. However, technical limitations in producing the large numbers of retroviral vectors required to infect a significant fraction of cells in the tissues of a human or vertebrate currently limit their use as gene therapy vectors. Another concern with retroviral vectors is that their random integration might disturb the normal expression of cellular genes encoding proteins regulating cellular replication. This type of cellular gene deregulation occurs naturally following infection with certain retroviruses, such as avian leukosis virus and murine leukemia viruses, leading to development of leukemia in birds and mice, respectively.

Adeno-associated virus (AAV) is a "satellite" parvovirus that replicates only in cells that are co-infected with adenovirus or herpes simplex virus. When AAV infects human cells in the absence of these "helper" viruses, its ssDNA genome is copied into dsDNA by host-cell DNA polymerase and then is integrated into a single region on chromosome 19, where it does not have any known deleterious effects. Research is under way to adapt the AAV integration mechanism that operates in the absence of helper virus to the development of a safe and effective integrating viral vector.

Chapter 7

Recombinant DNA and Genomics

Once the structure of DNA and the genetic code were unraveled, it became clear that many deep biological secrets were locked up in the sequence of bases in DNA. But identifying the sequences of long regions of DNA, much less altering them at will, seemed a distant dream. An avalanche of technical discoveries in the 1970s drastically changed this perspective and has led to astounding advances in molecular cell biology in the past two decades based on the analysis and manipulation of macromolecules, particularly DNA.

The discovery of two types of enzymes provided the impetus for these developments and permitted the now common technique of DNA cloning. One type, called restriction enzymes,cuts the DNA from any organism at specific sequences of a few nucleotides, generating a reproducible set of fragments. The other type, called DNA ligases, can insert DNA restriction fragments into replicating DNA molecules producingrecombinant DNA. The recombinant DNA molecules then can be introduced into appropriate cells, most often bacterial cells; all the descendants from a single such cell, called a clone,carry the same recombinant DNA molecule. Once a clone of cells bearing a desired segment of DNA is isolated, unlimited quantities of this DNA can be prepared. In addition, DNA sequences up to about 100 bases long can now be chemically synthesized by entirely automated procedures. Recombinant DNAs thus can be produced containing either natural DNA fragments resulting from restriction-enzyme cleavage or any desired chemically synthesized mutant sequences.

FROM PROTEIN TO GENE

isolate protein on the basis of its molecular function (e.g., enzymatic or hormonal activity)

↓

Determine partial amino acid sequence of the protein

↓

Synthesize oligonucleotides that correspond to portions of the amino acid sequece

↓

Use oligonucleotides as probes to select cDNA or genomic clone encoding the protein from library

↓

Sequence isolated gene

FROM GENE TO PROTEIN

isolate genomic clone corresponding to an altered trait in mutants (e.g., nutritional auxtrophy, inherited disease, developmental defect)

↓

Use genomic DA to isolate cDNA for the mRNA encoded by the gene

↓

Sequence the cDNA deduce amino acid sequence of the encoded protein

↓

Compare deduced amino acid sequence with that of known proteins to gain insight into function of the protein

↓

Use expression vector to produce the encoded protein

The availability of restriction enzymes also facilitated development of techniques for rapid DNA sequencing in the late 1970s. A long DNA molecule is first cleaved with restriction enzymes into a reproducible array of fragments, whose order in the original molecule is determined. Procedures also were developed for determining the sequence of bases in fragments up to 500 nucleotides long. Thus there was no longer any obstacle to obtaining the sequence of a DNA containing 10,000 or more nucleotides. Suddenly, any DNA could be isolated and sequenced. With the aid of computer-automated procedures for sequencing DNA and for storing, comparing, and analyzing sequence data, scientists will complete sequencing of the entire human genome in the next few years.

In the past, two basic approaches were available for unraveling the molecular basis of complex biological processes:

1. Biochemical purification and analysis of a protein based on its functional characteristics
2. Classical genetic analysis for the characterization and mapping of genes defined by mutations. The group of techniques discussed in this chapter, often collectively referred to as*recombinant DNA technology,*provide a link between these two types of experimental strategies, the analysis of proteins and the analysis of genes. Today's molecular cell biologists can begin with an isolated protein and clone the gene that encodes it. They also can reinsert cloned DNA, whether natural, modified, or completely synthetic, into cells and test its biological activity. Alternatively researchers can begin with the concept of a gene identified by the characteristics of a mutant organism and isolate a DNA clone containing the gene. Ultimately, the encoded protein can be produced in sufficient quantities for detailed study. The marriage of biochemical and genetic approaches by recombinant DNA technology provides an enormously powerful strategy for studying the role of particular proteins in cellular processes. In this chapter, we describe the various recombinant DNA techniques that permit this fruitful two-pronged approach, which is summarized in the flow diagram on the left.

DNA CLONING WITH PLASMID VECTORS

The essence of cell chemistry is to isolate a particular cellular component and then analyse its chemical structure and activity. In the case of DNA, this is feasible for relatively short molecules such as the genomes of small viruses. But genomes of even the simplest cells are much too large to directly analyse in detail at the molecular level. The problem is compounded for complex organisms. The human genome, for example, contains about 6×10^9 base pairs (bp) in the 23 pairs of chromosomes. Cleavage of human DNA with restriction enzymes that produce about one cut for every 3000 base pairs yields some 2 million fragments, far too many to separate from each other directly. This obstacle to obtaining pure DNA samples from large genomes has been overcome by recombinant DNA technology. With these methods virtually any gene can be purified, its sequence determined, and the functional regions of the sequence explored by altering it in planned ways and reintroducing the DNA into cells and into whole organisms.

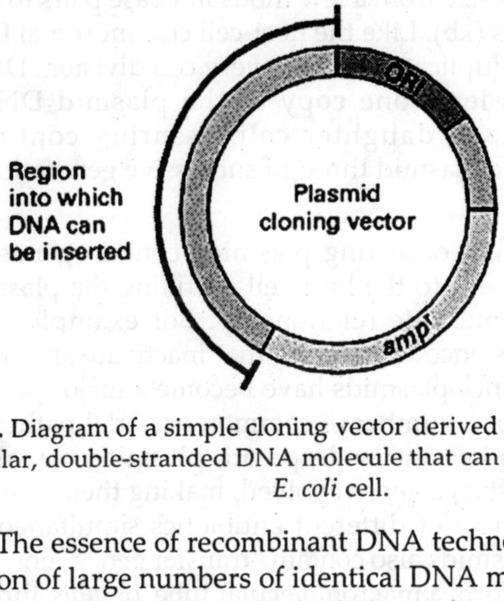

Fig. Diagram of a simple cloning vector derived from a plasmid, a circular, double-stranded DNA molecule that can replicate within an *E. coli* cell.

The essence of recombinant DNA technology is the preparation of large numbers of identical DNA molecules. A DNA fragment of interest is linked through standard 32 → 52 phosphodiester bonds to a vector DNA molecule, which can replicate when introduced into a host cell. When a single recombinant DNA molecule, composed of a vector plus an inserted DNA fragment, is introduced into a host cell, the inserted DNA is reproduced along with the vector, producing large numbers of recombinant DNA molecules that include the fragment of DNA originally linked to the vector. Two types of vectors are most commonly used: *E. coli*plasmid vectors and bacteriophage λ vectors. Plasmid vectors replicate along with their host cells, while λ vectors replicate as lytic viruses, killing the host cell and packaging the DNA into virions. In this section, the general procedure for cloning DNA fragments in *E. coli* plasmids is described.

PLASMIDS ARE EXTRACHROMOSOMAL SELF-REPLICATING DNA MOLECULES

Plasmids are circular, double-stranded DNA (dsDNA) molecules that are separate from a cell's chromosomal DNA. These extrachromosomal DNAs, which occur naturally in bacteria, yeast, and some higher eukaryotic cells, exist in a parasitic or symbiotic relationship with their host cell. Plasmids range in size from a few thousand base pairs to more than 100 kilobases (kb). Like the host-cell chromosomal DNA, plasmid DNA is duplicated before every cell division. During cell division, at least one copy of the plasmid DNA is segregated to each daughter cell, assuring continued propagation of the plasmid through successive generations of the host cell.

Many naturally occurring plasmids contain genes that provide some benefit to the host cell, fulfilling the plasmid's portion of the symbiotic relationship. For example, some bacterial plasmids encode enzymes that inactivate antibiotics. Such drug-resistance plasmids have become a major problem in the treatment of a number of common bacterial pathogens. As antibiotic use became widespread, plasmids containing several drug-resistance genes evolved, making their host cells resistant to a variety of different antibiotics simultaneously. Many of these plasmids also contain "transfer genes" encoding proteins that can form a macromolecular tube, or *pilus*, through which a copy of the plasmid can be transferred to other host cells of the same or related bacterial species. Such transfer can result in the rapid spread of drug-resistance plasmids, expanding the number of antibiotic-resistant bacteria in an environment such as a hospital. Coping with the spread of drug-resistance plasmids is an important challenge for modern medicine.

E. COLI PLASMIDS CAN BE ENGINEERED FOR USE AS CLONING VECTORS

The plasmids most commonly used in recombinant DNA

technology replicate in *E. coli*.Generally, these plasmids have been engineered to optimize their use as vectors in DNA cloning. For instance, to simplify working with plasmids, their length is reduced; many plasmid vectors are only ≈3kb in length, which is much shorter than in naturally occurring *E. coli* plasmids. (The circumference of plasmids usually is referred to as their "length," even though plasmids are almost always circular DNA molecules.) Most plasmid vectors contain little more than the essential nucleotide sequences required for their use in DNA cloning: a replication origin, a drug-resistance gene, and a region in which exogenous DNA fragments can be inserted.

PLASMID DNA REPLICATION

The replication origin (ORI) is a specific DNA sequence of 50 – 100 base pairs that must be present in a plasmid for it to replicate. Host-cell enzymes bind to ORI, initiating replication of the circular plasmid. Once DNA replication is initiated at ORI, it continues around the circular plasmid regardless of its nucleotide sequence. Thus any DNA sequence inserted into such a plasmid is replicated along with the rest of the plasmid DNA; this property is the basis of molecular DNA cloning.

SELECTION OF TRANSFORMED CELLS

In 1944, O. T. Avery, C. M. Macleod, and M. McCarty first demonstrated gene transfer with isolated DNA obtained from *Streptococcus pneumoniae.* This process involved the genetic alteration of a bacterial cell by the uptake of DNA isolated from a genetically different bacterium and its recombination with the host-cell genome. Their experiments provided the first evidence that DNA is the genetic material. Later studies showed that such genetic alteration of a recipient cell can result from the uptake of exogenous extrachromosomal DNA (e.g., plasmids) that does not integrate into the host-cell chromosome. The term transformation is used to denote the genetic alteration of a cell caused by the uptake and expression of foreign DNA regardless of the mechanism involved.

The phenomenon of transformation permits plasmid vectors to be introduced into and expressed by *E. coli* cells. In order to be useful in DNA cloning, however, a plasmid vector must contain a *selectable gene,* most commonly a drug-resistance gene encoding an enzyme that inactivates a specific antibiotic. As we've seen, the ampicillin-resistance gene (*amp*^{r}) encodes β-lactamase, which inactivates the antibiotic ampicillin. After plasmid vectors are incubated with *E. coli,* those cells that take up the plasmid can be easily selected from the larger number of cells that do not by growing them in an ampicillin-containing medium. The ability to select transformed cells is critical to DNA cloning by plasmid vector technology because the transformation of *E. coli* with isolated plasmid DNA is inefficient.

Normal *E. coli* cells cannot take up plasmid DNA from the medium. Exposure of cells to high concentrations of certain divalent cations, however, makes a small fraction of cells permeable to foreign DNA by a mechanism that is not understood. In a typical procedure, *E. coli* cells are treated with $CaCl_2$ and mixed with plasmid vectors; commonly, only 1 cell in about 10,000 or more cells becomes competent to take up the foreign DNA. Each competent cell incorporates a *single* plasmid DNA molecule, which carries an antibiotic-resistance gene. When the treated cells are plated on a petri dish of nutrient agar containing the antibiotic, only the rare transformed cells containing the antibiotic-resistance gene on the plasmid vector will survive. All the plasmids in such a colony of selected transformed cells are descended from the single plasmid taken up by the cell that established the colony.

PLASMID CLONING PERMITS ISOLATION OF DNA FRAGMENTS FROM COMPLEX MIXTURES

A DNA fragment of a few base pairs up to ≈20 kb can be inserted into a plasmid vector. When such a recombinant plasmid transforms an *E. coli* cell, all the antibiotic-resistant progeny cells that arise from the initial transformed cell will contain plasmids with the same inserted sequence of DNA. The inserted DNA is replicated along with the rest of the plasmid DNA and segregates to daughter cells as the colony grows. In this way, the initial fragment of DNA is replicated in the colony of cells into a large number of identical copies. Since all the cells in a colony arise from a single transformed parental cell, they constitute a clone of cells. The initial fragment of DNA inserted into the parental plasmid is referred to as *cloned DNA*, since it can be isolated from the clone of cells.

DNA cloning allows fragments of DNA with a particular nucleotide sequence to be isolated from a complex mixture of fragments with many different sequences. As a simple example, assume you have a solution containing four different types of DNA fragments, each with a unique sequence. Each fragment type is individually inserted into a plasmid vector. The resulting mixture of recombinant plasmids is incubated with *E. coli* cells under conditions that facilitate transformation; the cells then are cultured on antibiotic selective plates. Since each colony that develops arose from a single cell that took up a single plasmid, all the cells in a colony harbor the identical type of plasmid characterized by the DNA fragment inserted into it. As a result, copies of the DNA fragments in the initial mixture are isolated from one another in the separate bacterial colonies. DNA cloning thus is a powerful, yet simple method for purifying a particular DNA fragment from a complex mixture of fragments and producing large numbers of the fragment of interest.

RESTRICTION ENZYMES CUT DNA MOLECULES AT SPECIFIC SEQUENCES

To clone specific DNA fragments in a plasmid vector, as just described, or in other vectors discussed in later sections, the fragments must be produced and then inserted into the vector DNA. As noted in the introduction, restriction enzymes and DNA ligases are utilized to produce such recombinant DNA molecules.

Restriction enzymes are bacterial enzymes that recognize specific 4- to 8-bp sequences, called *restriction sites*, and then cleave both DNA strands at this site. Since these enzymes cleave DNA within the molecule, they are also called *restriction endonucleases* to distinguish them from exonucleases, which digest nucleic acids from an end. Many restriction sites are short inverted repeat sequences; that is, the restriction-site sequence is the same on each DNA strand when read in the 52 → 32 direction. Because the DNA isolated from an individual organism has a specific sequence, restriction enzymes cut the DNA into a reproducible set of fragments called restriction fragments.

The word *restriction* in the name of these enzymes refers to their function in the bacteria from which they are isolated: a restriction endonuclease destroys (restricts) incoming foreign DNA (e.g., bacteriophage DNA or DNA taken up during transformation) by cleaving it at all the restriction sites in the DNA. Another enzyme, called a *modification enzyme,* protects a bacterium's own DNA from cleavage by modifying it at or near each potential cleavage site. The modification enzyme adds a methyl group to one or two bases, usually within the restriction site. When a methyl group is present there, the restriction endonuclease is prevented from cutting the DNA. Together with the restriction endonuclease, the methylating enzyme forms a restriction-modification system that protects the host DNA while it destroys foreign DNA. Restriction enzymes have been purified from several hundred different species of bacteria, allowing DNA molecules to be cut at a large number of different sequences corresponding to the recognition sites of these enzymes.

RESTRICTION FRAGMENTS WITH COMPLEMENTARY "STICKY ENDS" ARE LIGATED EASILY

*Eco*RI makes staggered cuts in the two DNA strands. Many other restriction enzymes make similar cuts, generating fragments that have a single-stranded "tail" at both ends. The tails on the fragments generated at a given restriction site are complementary to those on all other fragments generated by the same restriction enzyme. At room temperature, these single-stranded regions, often called "sticky ends," can transiently base-pair with those on other DNA fragments generated with the same restriction enzyme, regardless of the source of the DNA. This base pairing of sticky ends permits DNA from widely differing species to be ligated, forming chimeric molecules.

During in vivo DNA replication, DNA ligase catalyzes formation of 32 → 52 phosphodiester bonds between the short fragments of the discontinuously synthesized DNA strand at a replication fork. In recombinant DNA technology, purified DNA ligase is used to covalently join the ends of restriction fragments in vitro. This enzyme can catalyze the formation of a 32 → 52 phosphodiester bond between the 32 -hydroxyl end of one restriction-fragment strand and the 52 -phosphate end of another restriction-fragment strand during the time that the sticky ends are transiently base-paired. When DNA ligase and ATP are added to a solution containing restriction fragments with sticky ends, the restriction fragments are covalently ligated together through the standard 32 → 52 phosphodiester bonds of DNA.

Some restriction enzymes, such as *Alu*I and *Sma*I, cleave both DNA strands at the same point within the recognition site. These restriction enzymes generate DNA restriction fragments with "blunt" (flush) ends in which all the nucleotides at the fragment ends are base-paired to nucleotides in the complementary strand. In addition to ligating complementary sticky ends, the DNA ligase from bacteriophage T4 can ligate any two blunt DNA ends. However, blunt-end ligation requires a higher DNA concentration than ligation of sticky ends.

POLYLINKERS FACILITATE INSERTION OF RESTRICTION FRAGMENTS INTO PLASMID VECTORS

Restriction enzymes to create fragments with sticky ends and DNA ligase to covalently link them allow foreign DNA to be inserted into plasmid vectors in vitro in a straightforward procedure. *E. coli* plasmid vectors can be constructed with a *polylinker*, a synthetic multiple-cloning-site sequence that contains one copy of several different restriction sites. When such a vector is treated with a restriction enzyme that recognizes a recognition sequence in the polylinker, it is cut at that sequence, generating sticky ends. In the presence of DNA ligase, DNA fragments produced with the same restriction enzyme will be inserted into the plasmid. The ratio of DNA fragments to be inserted to cut vectors and other reaction conditions are chosen to maximize the insertion of one restriction fragment per plasmid vector. The recombinant plasmids produced in in vitro ligation reactions then can be used to transform antibiotic-sensitive *E. coli* cells. All the cells in each antibiotic-resistant clone that remains after selection contain plasmids with the same inserted DNA fragment, but different clones carry different fragments.

SMALL DNA MOLECULES CAN BE CHEMICALLY SYNTHESIZED

Advances in synthetic chemistry now permit the chemical synthesis of single-stranded DNA (ssDNA) molecules of any sequence up to about 100 nucleotides in length. Synthetic DNA has a number of applications in recombinant DNA technology. Complementary ssDNAs can be synthesized and hybridized to each other to form a dsDNA with sticky ends. Such completely synthetic dsDNAs can be cloned into plasmid vectors just as DNA restriction fragments prepared from living organisms are. For example, the 57-bp polylinker sequence was chemically synthesized and then inserted into plasmid vectors to facilitate the cloning of fragments generated by different restriction enzymes. This example illustrates the use of synthetic DNAs to add convenient restriction sites where they otherwise do not occur. As described later in the chapter, synthetic DNAs are used in sequencing DNA and as probes to identify clones of interest. Synthetic DNAs also can be substituted for natural DNA sequences in cloned DNA to study the effects of specific mutations.

The technique for chemical synthesis of DNA oligonucleotides. Note that chains grow in the 32 → 52 direction, opposite to the direction of DNA chain growth catalyzed by DNA polymerases. Once the chemistry for producing synthetic DNA was standardized, automated instruments were developed that allow researchers to programme the synthesis of oligonucleotides of specific sequences up to about 100 nucleotides long.

CONSTRUCTING DNA LIBRARIES WITH λ PHAGE AND OTHER CLONING VECTORS

Most DNA cloning is done with *E. coli* plasmid vectors because of the relative simplicity of the cloning procedure. However, the number of individual clones that can be obtained by plasmid cloning is limited by the relatively low efficiency of *E. coli* transformation and the small number (only a few hundred) of individual transformed colonies that can be grown on a typical culture plate. These limitations make plasmid cloning of all the genomic DNA of higher organisms impractical. For example, $\approx 1.5 \times 10^5$ clones carrying 20-kb DNA fragments are required to represent the total human haploid genome, which contains $\approx 3 \times 10^9$ base pairs. Fortunately, cloning vectors derived from various bacteriophages have proved to be a practical means for obtaining the required number of clones to represent large genomes. A collection of clones that includes *all* the DNA sequences of a given species is called a *genomic DNA library,* or simply *genomic library.* Once a genomic library is prepared, it can be screened for clones containing a sequence of interest.

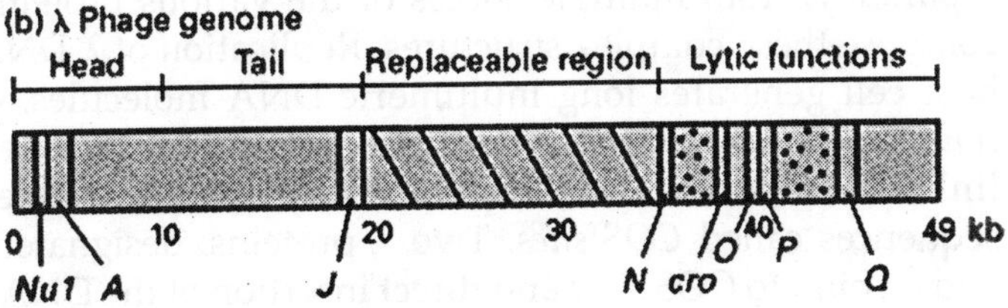

Fig. The bacteriophage genome.

BACTERIOPHAGE λ CAN BE MODIFIED FOR USE AS A CLONING VECTOR AND ASSEMBLED IN VITRO

Bacteriophage λ is probably the most extensively studied bacterial virus, and a great deal is known about its molecular biology and genetics. A λ phage virion has a head, which contains the viral DNA genome, and a tail, which functions in infecting *E. coli* host cells. When λ DNA enters the host-cell cytoplasm following infection, it undergoes either lytic or lysogenic growth. In lytic growth, the viral DNA is replicated and assembled into more than 100 progeny virions in each infected cell, killing the cell in the process and releasing the replicated virions. In lysogenic growth, the viral DNA inserts into the bacterial chromosome, where it is passively replicated along with the host-cell chromosome as the cell grows and divides.

The λ genes encoding the head and tail proteins as well as various proteins involved in the lytic and lysogenic growth pathways are clustered in discrete regions of the ≈50-kb viral genome. When bacteriophage λ is used as a cloning vector, it must be capable of lytic growth, but other viral functions are irrelevant. Consequently, the genes involved in the lysogenic pathway and other viral genes not essential for the lytic pathway are removed from the viral DNA and replaced with the DNA to be cloned. Up to ≈25 kb of foreign DNA can be inserted into the λ genome, resulting in a recombinant DNA that can be packaged in vitro to form virions capable of replicating and forming plaques on *E. coli* host cells.

During the in vivo assembly of λ virions within infected host cells, viral heads and tails initially are assembled separately, from multiple copies of the various proteins that compose these complex structures. Replication of λ DNA in a host cell generates long multimeric DNA molecules, called *concatomers,* that consist of multiple copies of the viral genome linked end to end and separated by specific nucleotide sequences called *COS sites.* Two λ proteins, designated Nu1 and A, bind to COS sites and direct insertion of the DNA lying between two adjacent COS sites into a preassembled head. This process results in the packaging of a single ≈50-kb λ genome from the multimeric concatomer into each preassembled head. Host-cell chromosomal DNA is not inserted into the λ heads because it does not contain any copies of the COS sequence. Once λ DNA is inserted into a preassembled λ head, the preassembled tail is attached, producing a complete virion.

To prepare infectious λ virions carrying recombinant DNA, the phage-assembly process is carried out in vitro. In one method, *E. coli* cells are infected with a λ mutant defective in A protein, one of the two proteins required for packaging λ DNA into preassembled phage heads. These cells accumulate preassembled "empty" heads; since tails attach only to heads "filled" with DNA. Preassembled tails also accumulate in these cells. An extract containing high concentrations of empty heads and tails is prepared by lysing cells infected with the λ A mutant. When this extract is mixed with isolated A protein (obtained from λ-infected cells) and recombinant λ DNA containing a COS site, the DNA is packaged into the empty heads. The tails in the extract then combine with the filled heads, yielding complete virions carrying the recombinant λ DNA.

This procedure produces a high yield of recombinant virions that are fully infectious and can efficiently infect *E. coli* cells. Each virion particle binds to receptors on the surface of a host cell and injects its packaged recombinant DNA into the cell. Infection by λ phage is about a thousand times more efficient than transformation with plasmid vectors, accounting for the high efficiency of λ phage cloning. For instance, $\approx 10^6$ transformed colonies per microgram of recombinant plasmid DNA can be obtained routinely, whereas $\approx 10^9$ plaques representing λ clones can be obtained per microgram of recombinant λ DNA.

NEARLY COMPLETE GENOMIC LIBRARIES OF HIGHER ORGANISMS CAN BE PREPARED BY λ CLONING

With the availability of λ phage cloning vectors, preparation of genomic libraries for higher organisms, including humans, is feasible. A genomic library is a set of λ (or plasmid) clones that collectively contain every DNA sequence in the genome of a particular organism. The λ DNA first is treated with a restriction enzyme to produce fragments called λ *vector arms,* which have sticky ends and together contain all the genes necessary for lytic growth. This step frees the nonessential region in the middle of the λ genome; this region is separated from the λ arms and discarded. Genomic DNA then is extracted from a cell type that contains all the genetic information of the organism under study. Sperm cells or cells of an early embryo often are used as sources of mammalian DNA. The extracted DNA then is cleaved by a restriction enzyme to produce ≈20-kb fragments with sticky ends complementary to the sticky ends on the λ vector arms being used.

The λ arms and the collection of genomic DNA fragments are mixed in about equal amounts. The complementary sticky ends on the fragments and λ arms hybridize and then are joined covalently by DNA ligase. Each of the resulting recombinant DNA molecules contains a foreign DNA fragment located between the two arms of the λ vector DNA. The ligated recombinant DNAs then are packaged into λ virions in vitro as described above. Only DNA molecules of the correct size can be packaged to produce fully infectious recombinant λ virions.

Finally, the recombinant λ virions are plated on a lawn of *E. coli* cells to generate a large number of recombinant λ plaques. Since each plaque arises from a single recombinant virion, all the progeny λ phages that develop are genetically identical and constitute a clone carrying a particular genomic DNA insert. The different plaques correspond to distinct phage clones, each carrying a different DNA insert, and collectively they constitute a λ genomic library. Multiple λ vectors have been constructed containing different restriction sites, so that restriction fragments generated by a variety of restriction enzymes can be cloned in λ vectors.

In constructing a library of genomic DNA, the DNA commonly is cleaved with a restriction enzyme that recognizes a 4-bp restriction site. A specific 4-bp sequence will occur on average once every $4^4 = 256$ base pairs. Complete digestion of the human haploid genome, which contains $\approx 3 \times 10^9$ base pairs, would yield somewhat more than 10^7 nonoverlapping different fragments. However, to increase the probability that all regions of the genome are successfully cloned and represented in the λ genomic library, the genomic DNA usually is only partially digested to yield overlapping restriction fragments of ≈20 kb. In a large λ library constructed from such overlapping restriction fragments, a specific sequence of genomic DNA (e.g., the gene encoding β-globin) often is contained in or extends over several "overlapping" clones. By a technique called chromosome walking, described in the next chapter, overlapping genomic clones can be ordered.

The size of a genomic library for a given organism depends on the amount of DNA in that organism's haploid genome. If the human genome of about 3×10^9 base pairs is cleaved into 20-kb fragments for insertion into a λ vector, then roughly 1.5×10^5 different recombinant λ phage virions would be required to constitute a complete library. Because the restriction fragments of human DNA are incorporated into phages randomly, about 10^6 recombinant phages are necessary to assure that each region of human DNA has a 90 – 95 percent chance of being included.

Each plaque produced by a recombinant bacteriophage λ contains large numbers of recombinant virions and, consequently, large numbers of identical cloned DNA fragments. Hybridization methods for identifying recombinant λ clones of interest are described in a later section. Because these methods allow specific detection of very small plaques, as many as 5×10^4 plaques can be screened on a single culture plate. Thus only 20 – 30 petri dishes, each containing about 5×10^4 λ plaques, are sufficient to represent the entire human genome. In contrast, to screen 10^6 recombinant plasmids carrying the entire human genome would require about 5000 petri dishes, because only 200 or so transformed *E. coli* colonies can be detected on a typical petri dish.

CDNA LIBRARIES ARE PREPARED FROM ISOLATED MRNAS

In higher eukaryotes, many genes are transcribed into mRNA only in specialized cell types. For example, mRNAs encoding globin proteins are found only in erythrocyte precursor cells, called *reticulocytes.* Likewise, the mRNA encoding albumin, the major protein in serum, is produced only in liver cells where albumin is synthesized. The specific DNA sequences expressed as mRNAs in a particular cell type can be cloned by synthesizing DNA copies of the mRNAs isolated from that type of cell, and then cloning the DNA copies in plasmid or bacteriophage λ vectors.

DNA copies of mRNAs are called complementary DNAs(cDNAs); clones of such DNA copies of mRNAs are called *cDNA clones.* In addition to representing only the sequences expressed as mRNAs in a particular cell type, cDNA clones lack the noncoding introns present in genomic DNA clones. Thus the amino acid sequence of a protein can be determined directly from the nucleotide sequence of its corresponding cDNA. Many genes in higher eukaryotes are too large to be included in a single λ clone because of their large introns. In contrast, all full-length cDNAs, containing the entire protein-coding sequence, can be included in a single λ clone. However, because of methodological difficulties, not all cDNA clones are full length when initially produced; to obtain a full-length cDNA, it often is necessary to isolate several overlapping cDNA clones and then ligate them at rare restriction sites. Just as a large collection of clones containing fragments of genomic DNA representing the entire genome of a species is called a *genomic library,* a large collection of cDNA copies of all the mRNAs in a cell type is called a *cDNA library.*

ISOLATION OF MRNAS AND SYNTHESIS OF CDNAS

The first step in preparing a cDNA library is to isolate the total mRNA from the cell type or tissue of interest. Nature has greatly simplified the isolation of eukaryotic mRNAs: the 32 end of nearly all eukaryotic mRNAs consists of a string of

50 – 250 adenylate residues, the *poly(A) tail.* Because of their poly(A) tail, mRNAs can be easily separated from the much more prevalent rRNAs and tRNAs present in a cell extract by use of a column to which short strings of thymidylate (oligo-dTs) are linked to the matrix. When a cell extract is passed through an oligo-dT column, the mRNA poly(A) tails base-pair with the oligo-dTs, binding the mRNAs to the column. Since rRNAs, tRNAs, and other molecules do not bind to the column, they can be washed away. The bound mRNAs are recovered by elution with a low-salt buffer.

The enzyme reverse transcriptase, which is found in retroviruses, is then used to synthesize a strand of DNA complementary to each mRNA molecule. This enzyme can polymerize deoxynucleoside triphosphates into a complementary DNA strand using an RNA molecule as template. Like other DNA polymerases, reverse transcriptase can add nucleo-tides only to the 32 end of a preexisting primer base-paired to the template. Added free oligo-dT serves this function by hybridizing to the 32 poly(A) tail of each mRNA template.

CONVERSION OF SINGLE-STRANDED CDNA TO DOUBLE-STRANDED CDNA

After cDNA copies of isolated mRNAs are synthesized, the mRNAs are removed by treatment with alkali, which hydrolyzes RNA but not DNA. The single-stranded cDNAs then are converted to double-stranded DNA molecules. To do this, the 32 end of each cDNA strand is elongated by adding several residues of a single nucleotide (e.g., dG) through the action of *terminal transferase,* a unique DNA polymerase that does not require a template, but simply adds deoxynucleotides to free 32 ends. A synthetic oligo-dC primer then is hybridized to this 32 oligo-dG. DNA polymerase, which uses the oligo-dC as a primer, then is used to synthesize a DNA strand complementary to the original cDNA strand. These reactions produce a complete double-stranded DNA molecule corresponding to each of the mRNA molecules in the original preparation Each double-stranded DNA, also called cDNA, contains an oligo-dC – oligo-dG double-stranded region at one end and an oligo-dT – oligo-dA double-stranded region at the other end.

ADDITION OF LINKERS AND INCORPORATION OF CDNA INTO A VECTOR

To prepare double-stranded cDNAs for cloning, short *restriction-site linkers* first are ligated to both ends. These are double-stranded DNA segments, usually ≈10 – 12 bp long, that contain the recognition site for a particular restriction enzyme. Restriction-site linkers are prepared by hybridizing chemically synthesized complementary oligonucleotides. The ligation reaction is carried out by DNA ligase from bacteriophage T4, which can join "blunt-ended" double-stranded DNA molecules lacking sticky ends. Although blunt-end ligation is relatively inefficient, the ligation reaction can be driven to completion by using high concentrations of linkers.

The resulting double-stranded cDNAs, which contain a restriction-site linker at each end, are treated with the restriction enzyme specific for the linker; this generates cDNA molecules with sticky ends at each end. To prevent digestion of any cDNAs that by chance have a recognition sequence for this restriction enzyme within the cDNA sequence, the mixture of double-stranded cDNAs is treated with the appropriate modification enzyme before addition of the linkers. This enzyme methylates specific bases within the restriction-site sequence, preventing the restriction enzyme from digesting the methylated sites.

The final step in construction of a cDNA library is ligation of the restriction-cleaved double-stranded cDNAs, which now have sticky ends, to plasmid or λ phage vectors that have been cut to generate complementary sticky ends. The recombinant vectors then are plated on a lawn of *E. coli* cells, producing a library of plasmid or λ clones. Each clone carries a cDNA derived from a single mRNA.

LARGER DNA FRAGMENTS CAN BE CLONED IN COSMIDS AND OTHER VECTORS

Both λ phage vectors and the more commonly used *E. coli* plasmid vectors are useful for cloning DNA fragments up to

≈20 –†25 kb. However, cloning of much larger fragments is desirable for sequencing of extremely long DNAs such as the DNA in a eukaryotic chromosome. Also, because of the common occurrence of large introns in genes from higher eukaryotes, it is often necessary to clone DNA fragments greater than 25 kb in order to include an entire gene in one clone. Consequently, additional types of cloning vectors have been developed for cloning larger fragments of DNA.

One common method for cloning larger fragments makes use of elements of both plasmid and λ phage cloning. In this method, called *cosmid cloning*, recombinant plasmids containing inserted fragments up to 45 kb long can be efficiently introduced into *E. coli* cells. A cosmid vector is produced by inserting the COS sequence from λ phage DNA into a small *E. coli* plasmid vector about 5 kb long. Like other plasmid vectors discussed earlier, cosmid vectors contain a replication origin (ORI), an antibiotic-resistance gene (e.g., amp^r), and a polylinker sequence containing numerous restriction-enzyme recognition sites. Next, the cosmid vector is cut with a restriction enzyme and then ligated to 35- to 45-kb restriction fragments of foreign DNA with complementary sticky ends. If the concentration of foreign DNA is high enough, the ligation reaction generates long DNA molecules containing multiple restriction fragments of the foreign DNA separated by the 5-kb cosmid DNA. These ligated DNA molecules, which resemble the concatomers that form during replication of λ phage in a host cell, can be packaged in vitro as described earlier.

In the packaging reaction, the λ Nu1 and A proteins bind to COS sites in the ligated DNA and direct insertion of the DNA between two adjacent COS sites into empty phage heads. Packaging will occur so long as the distance between adjacent COS sites does not exceed about 50 kb (the approximate size of the λ genome). Phage tails then are attached to the filled heads, producing viral particles that contain a recombinant cosmid DNA molecule rather than the λ genome. When these virions are plated on a lawn of *E. coli* cells, they bind to phage receptors on the cell surface and inject the packaged DNA into the cells.

Since the injected DNA does not encode any λ proteins, no viral particles form in infected cells and no plaques develop on the plate. Rather, the injected DNA forms a large circular plasmid, composed of the cosmid vector and an inserted DNA fragment, in each host cell. This plasmid replicates and is segregated to daughter cells like other *E. coli* plasmids and the colonies that arise from transformed cells can be selected on antibiotic plates. The high efficiency of λ phage infection of *E. coli* cells makes cosmid cloning a practical method of generating plasmid clones carrying DNA fragments up to 45 kb long. Since many genes of higher eukaryotes are on the order of 30 –†40 kb in length, cosmid cloning increases the chances of obtaining DNA clones containing the entire sequences of genes.

Another strategy similar to cosmid cloning makes use of larger *E. coli* viruses such as bacteriophage P1, whose head can accommodate larger DNA molecules than the λ phage head. Recombinant plasmids containing DNA fragments up to ≈100 kb long can be packaged in vitro with the P1 system.

Still larger fragments, up to 300 kb long, can be cloned in *bacterial artificial chromosomes* (BACs). These vectors are similar to standard *E. coli* plasmid vectors except that they contain the origin and genes encoding the ORIbinding proteins required for plasmid replication from a naturally occurring large *E. coli* plasmid called the *F-factor*. Recombinant BACs containing large DNA fragments are produced as in standard plasmid vector cloning. However, an alternative procedure for introducing DNA into cells, called *electroporation,* is used to overcome the very low efficiency of *E. coli* transformation by large plasmids. In this procedure, cells are mixed with DNA and subjected to a brief pulse of extremely high voltage. By an unknown mechanism, DNA molecules as long as 300 kb can be introduced into *E. coli* cells at a sufficiently high frequency to make BAC cloning practical. Large fragments cloned in BACs have been found to be stable through hundreds of generations.

Even larger fragments of DNA, containing up to 1000 kb, can be cloned into *yeast artificial chromosomes* (YACs). To understand how YACs function requires an explanation of the basic elements of eukaryotic chromosomes.

IDENTIFYING, ANALYZING, AND SEQUENCING CLONED DNA

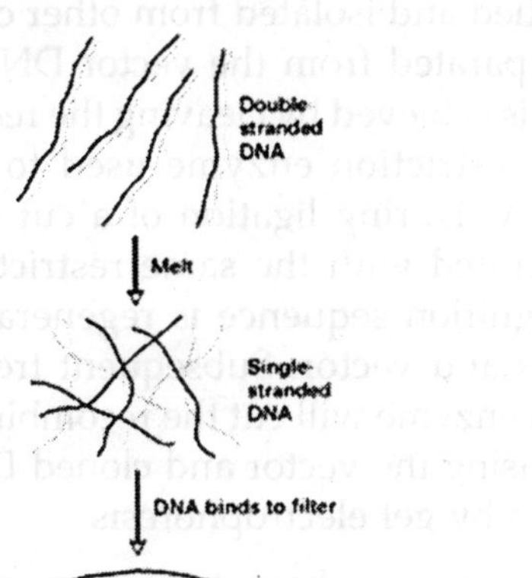

Fig. Membrane-hybridization assay for detecting nucleic acids.

Suppose you have isolated a particular protein and want to isolate the gene that encodes it. A complete genomic λ library from mammals contains at least a million different clones; a cDNA library must contain as many clones to include the sequences of scarce mRNAs. How are specific clones of interest identified in such large collections? The most common method involves screening a library by hybridization with radioactively labeled DNA or RNA probes. In an alternative method, a specific clone in a library of cloned DNA is identified based on some property of its encoded protein.

Once a particular genomic DNA or cDNA clone of interest has been identified and isolated from other clones, the cloned DNA can be separated from the vector DNA and analyzed. This separation is achieved by cleaving the recombinant vector with the same restriction enzyme used to insert the DNA fragment initially. During ligation of a cut vector and DNA fragments generated with the same restriction enzyme, the restriction recognition sequence is regenerated between the DNA fragments and vector. Subsequent treatment with the same restriction enzyme will cut the recombinant vector at the same sites, releasing the vector and cloned DNA, which then can be separated by gel electrophoresis.

The most complete characterization of a cloned DNA requires determination of its nucleotide sequence; from this sequence, the amino acid sequence of an encoded protein can be deduced. The sequence of genomic DNA includes introns as well as exons; it also includes regions that control gene expression by determining the type of cell in which the encoded protein is expressed, the stage in development it is expressed, and the amount of protein produced. Genomic DNAs also include replication origins and sequences important in determining how the DNA associates with proteins in chromosomes. In subsequent chapters, we consider how cells use DNA sequences for these functions. In this section, techniques for identifying, characterizing, and finally sequencing cloned DNA are outlined.

LIBRARIES CAN BE SCREENED WITH MEMBRANE-HYBRIDIZATION ASSAY

As discussed under the conditions of temperature and ion concentration found in cells, DNA is maintained as a duplex (double-stranded) structure by the hydrogen bonds between A · T and G · C base pairs. DNA duplexes can be denatured (melted) into single strands by heating them in a dilute salt solution (e.g., 0.01 M NaCl), or by raising the pH above 11. If the temperature is lowered and the ion concentration in the solution is raised, or if the pH is lowered to neutrality, the A · T and G · C base pairs re-form between complementary single strands. This process goes by many names: renaturation, reassociation, hybridization, annealing. In a mixture of nucleic acids, only complementary single strands (or strands containing complementary regions) will reassociate; the extent of their reassociation is virtually unaffected by the presence of noncomplementary strands. Such *molecular hybridization* can take place between two complementary strands of either DNA or RNA, or between an RNA strand and a DNA strand.

To detect specific DNA clones by molecular hybridization, cloned recombinant DNA molecules are denatured and the single strands attached to a solid support, commonly a nitrocellulose filter or treated nylon membrane. When a solution containing single-stranded nucleic acids is dried on such a membrane, the single strands become irreversibly bound to the solid support in a manner that leaves most of the bases available for hybridizing to a complementary strand. Although the chemistry of this irreversible binding is not well understood, the procedure is very useful. The membrane is then incubated in a solution containing a radioactively labeled single-stranded DNA (or RNA) *probe* that is complementary to some of the nucleic acid bound to the membrane. Under hybridization conditions (near neutral pH, 40 – 65 °C, 0.3 – 0.6 M NaCl), this labeled probe hybridizes to the complementary nucleic acid bound to the membrane. Any excess probe that does not hybridize is washed away, and the labeled hybrids are detected by autoradiog-raphy of the filter.

The recombinant λ virions present in plaques on a lawn of *E. coli* are transferred to a nylon membrane by placing the membrane on the surface of the petri dish. Many of the viral particles in each plaque adsorb to the surface of the membrane, but many virions remain in the plaques on the surface of the nutrient agar in the petri dish. In this way a *replica* of the petri dish containing a large number of individual λ clones is reproduced on the surface of the membrane. The original petri dish is refrigerated to store the collection of λ clones. The membrane is then incubated in an alkaline solution, which disrupts the virions, releasing and denaturing the encapsulated DNA. The membrane is then dried, fixing the recombinant λ DNA to the membrane's surface. Next, the membrane is incubated with a radiolabeled probe under hybridization conditions. Unhybridized probe is washed away, and the filter is subjected to autoradiography.

The appearance of a spot on the autoradiogram indicates the presence of a recombinant λ clone containing DNA complementary to the probe. The position of the spot on the autoradiogram corresponds to the position on the original petri dish where that particular clone formed a plaque. Since the original petri dish still contains many infectious virions in each plaque, viral particles from the identified clone can be recovered for replating by aligning the autoradiogram and the petri dish and removing viral particles from the clone corresponding to the spot. A similar technique can be applied for screening a plasmid library in *E. coli* cells.

OLIGONUCLEOTIDE PROBES ARE DESIGNED BASED ON PARTIAL PROTEIN SEQUENCES

Identification of specific clones by the membrane-hybridization technique depends on the ability to prepare specific, radiolabeled probes. Specific oligonucleotide probes for the gene encoding a protein of interest can be synthesized chemically if a portion of the amino acid sequence of the protein is determined. For an oligonucleotide to be useful as a probe, it must be long enough for its sequence to occur uniquely in the clone of interest and not in any other clones. For most purposes, this condition is satisfied by oligonucleotides containing about 20 nucleotides. This is because a specific 20-nucleotide sequence occurs once in every 4^{20} ($\approx 10^{12}$) nucleotides. Since all genomes are much smaller ($\approx 3 \times 10^9$ nucleotides for humans), a specific 20-nucleotide sequence in a genome usually occurs only one time. In principle, probes this length can be prepared based on only a 7-aa sequence out of a protein's total sequence (20 nucleotides ÷ 3 nucleotides per codon ≈ 7 amino acids). However, a somewhat longer amino acid sequence usually is determined to allow the preparation of several probes for use in cloning a gene of interest.

Here we describe two different approaches for preparing oligonucleotide probes. Generally, a radiolabeled oligonucleotide probe is used to screen a λ cDNA library using the membrane-hybridization technique. Once a cDNA clone encoding a particular protein is obtained, the full-length radiolabeled cDNA can be used to probe a genomic library for clones that contain fragments of the gene encoding the protein.

DEGENERATE PROBES

The purified protein of interest is digested with one or more proteases (e.g., trypsin) into specific peptides, and the N-terminal amino acid sequences of a few of these peptides is determined by sequential Edman degradation or mass spectrometry. Based on the genetic code, the oligonucleotide sequences encoding the determined peptide sequences can be predicted. Recall, however, that the genetic code is degenerate; that is, many amino acids are encoded by multiple codons. Since the specific codons used to encode the protein of interest are unknown, oligonucleotides containing all possible combinations of codons must be synthesized to assure that one of them will match the gene perfectly.

Once several peptides have been sequenced, the 6- or 7-aa stretch that can be encoded by the smallest number of possible DNA sequences is determined. For example the amino acids in the sequence extending from position 3 through 8 (Cys-Ile-Tyr-Met-His-Gln) can be encoded by 2, 3, 2, 1, 2, and 2 possible codons, respectively. Consequently, 48 (= 2 × 2 × 1 × 2 × 2) different 18-base DNA sequences could encode this one sequence of amino acids. The GA added at the 32 end of these 18-base sequences must be complementary to the gene since the next amino acid in this peptide, Asp-9, is encoded by two codons that both start with GA. To be certain of obtaining a probe based on this amino acid sequence that hybridizes perfectly to the unique sequence present in the gene, all 48 of the 20-mer probes must be synthesized. A mixture of 20-mer probes based on any other portion of this peptide sequence would have to contain considerably more than 48 oligonucleotides because of the presence of leucine or serine residues, each encoded by six different codons.

A mixture of all the oligonucleotides that can encode a selected portion of a peptide sequence is called a *degenerate probe.* Such a mixture can be prepared at one time by adding more than one nucleotide precursor to the synthesis reaction at those points in the sequence that can be encoded by alternative bases. The final step in preparing this type of probe is to radiolabel the oligonucleotides, usually by transferring a ^{32}P-labeled phosphate group from ATP to the 52 end of each oligonucleotide using *polynucleotide kinase.* Screening of a λ cDNA library with a degenerate probe using the membrane-hybridization technique will identify clones that hybridize to the perfectly complementary oligonucleotide present in the probe mixture. Under the usual experimental conditions, oligonucleotides that differ from the cDNA sequence at one or two bases also will hybridize.

UNIQUE EST-BASED PROBES

In recent years another approach has become available for obtaining a probe based on the partial amino acid sequence determined from an isolated protein. Because this approach utilizes cDNA sequence data, it identifies a single oligonucleotide, rather than a degenerate mixture, that can be used to screen a library for a particular gene. Using methods for DNA sequencing described later, researchers have sequenced portions of vast numbers of cDNAs isolated from human cells and some additional important model organisms such as the mouse, *Drosophila,* and the roundworm *Caenorhabditis elegans.* These partial cDNA sequences, generally 200 to 400 bp in length, have been stored in computers and are available to researchers throughout the world via the

Internet, the international computer network. This collection of partial cDNA sequences is called the *expressed sequence tag (EST)* database because it is composed of relatively short portions (tags) of genomic DNA sequence that are expressed in the form of mRNA. The EST database is constantly updated as sequences from increasing numbers of cDNA clones are added to it.

Computer programmes that apply the genetic code are used to translate the EST sequences into partial protein amino acid sequences. Using programmes that have been developed for the purpose and a personal computer, a researcher can search the current EST database for an EST that encodes a specific partial amino acid sequence in the particular protein under study. If a match is found, then the EST provides the unique DNA sequence of that portion of the full-length cDNA. A single, specific probe up to ≈100 bases long that is perfectly complementary to a portion of the EST can then be synthesized and radiolabeled. Alternatively, the polymerase chain reaction, described at the end of the chapter, can be used to synthesize a probe equal to the full length of the EST. By now, the human EST database is so large that ESTs can be identified that encode partial amino acid sequences determined from most isolated human proteins.

SPECIFIC CLONES CAN BE IDENTIFIED BASED ON PROPERTIES OF THE ENCODED PROTEINS

Genomic and cDNA libraries can also be screened for the properties of a specific protein encoded in the cloned DNA. This approach uses special cloning vectors, called λ expression vectors, in which the cloned DNA is transcribed into mRNA, which in turn is translated into the encoded protein. For example, λ phage vectors have been constructed so that the junction of inserted DNA lies in a region of the vector that is transcribed and translated at a high rate. Cloned DNA inserted at this position is transcribed into mRNA in every cell infected by this type of vector. If the cloned DNA contains a protein-coding sequence inserted in the same reading frame as the vector protein, infected cells will produce a *fusion protein* in which the amino terminus is encoded by the vector DNA and the remainder of the molecule by the cloned DNA.

When replica nitrocellulose filters are prepared from a recombinant library constructed in a λ expression vector, fusion proteins expressed from each individual clone are bound to the nitrocellulose filter. The replica filter can be screened by procedures capable of detecting specific fusion proteins. For example, a monoclonal antibody specific for a protein of interest can be incubated with replica filters of a λ cDNA expression library. If one of the λ clones expresses a fusion protein that includes the region of the protein bound by that monoclonal antibody, antibody molecules will bind to the filter at the position of that specific clone. After washing of the filter to remove unbound antibody, the position of the specific clone is detected by incubation with a second radioactively labeled antibody that recognizes the first antibody, followed by autoradiography of the filter.

In this method, termed expression cloning, any molecule that binds to a protein of interest with high affinity and specificity can be labeled and used as a probe to identify clones expressing the interacting protein. For instance, expression cloning has been useful in identifying cDNA clones encoding proteins that bind to specific DNA sequences; many such proteins are involved in controlling transcription. In this case, a labeled synthetic double-stranded DNA probe is incubated with replica filters prepared from a cDNA library cloned into a λ expression vector. Binding of the labeled DNA by fusion proteins locates the positions of desired clones on the original filter. As described in a later section, other types of expression vectors can be used to produce large amounts of a protein from a cloned gene.

GEL ELECTROPHORESIS RESOLVES DNA FRAGMENTS OF DIFFERENT SIZE

Once a specific DNA clone has been isolated, the cloned DNA is separated from the vector DNA by cleavage with the restriction enzyme used to form the recombinant plasmid, as described earlier. The cloned DNA and vector DNA then are separated by gel electrophoresis, a powerful method for separating proteins according to size. Gel electrophoresis also is used to separate DNA and RNA molecules by size and to estimate the size of nucleic acid molecules of unknown length by comparison with the migration of molecules of known length.

DNA and RNA molecules are highly charged near neutral pH because the phosphate group in each nucleotide contributes one negative charge. As a result, DNA and RNA molecules move toward the positive electrode during gel electrophoresis. Smaller molecules move through the gel matrix more readily than larger molecules, so that molecules of different length, such as restriction fragments, separate. Because the gel matrix restricts random diffusion of the molecules, molecules of different length separate into "bands" whose width equals that of the well into which the original DNA mixture was placed. The resolving power of gel electrophoresis is so great that single-stranded DNA molecules up to about 500 nucleotides long can be separated if they differ in length by only 1 nucleotide. This high resolution is critical to the DNA-sequencing procedures described later. DNA molecules composed of up to ≈2000 nucleotides usually are separated electrophoretically on *polyacrylamide gels,* and molecules from 500 nucleotides to 20 kb on *agarose gels.*

Two methods are common for visualizing separated DNA bands on a gel. If the DNA is not radiolabeled, the gel is incubated in a solution containing the fluorescent dye ethidium:

Ethidium

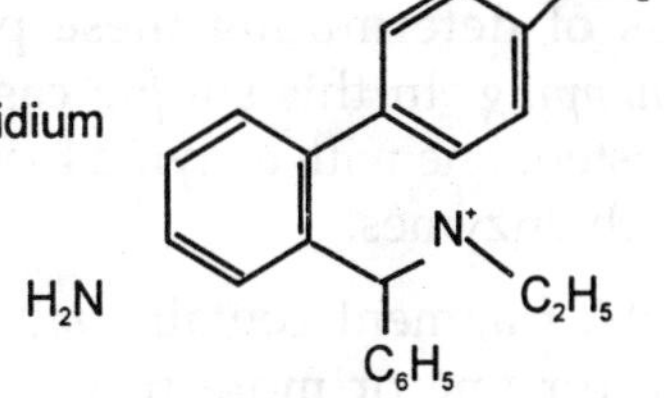

This planar molecule binds to DNA by intercalating between the base pairs. Binding concentrates ethidium in the DNA and also increases its intrinsic fluorescence. As a result, when the gel is illuminated with ultraviolet light, the regions of the gel containing DNA fluoresce much more brightly than the regions of the gel without DNA.

Radioactively labeled DNA can be visualized by autoradiography of the gel. In this case, the gel is laid against a sheet of photographic film in the dark, exposing the film at the positions where labeled DNA is present. When the film is developed, a photographic image of the DNA is observed. Radiolabeled DNA bands also can be detected by laying the gel against a phosphorimager screen, which counts β particles released by labeled molecules in the gel. The resulting data is stored by a computer and can be converted into an image of the gel that looks much like an autoradiogram.

MULTIPLE RESTRICTION SITES CAN BE MAPPED ON A CLONED DNA FRAGMENT

In addition to separating restriction fragments of different lengths, gel electrophoresis provides a means for estimating the length of fragments. As we've seen, the distance that a restriction fragment migrates in a gel is inversely proportional to the logarithm of its length. Thus the length of a restriction fragment can be determined fairly accurately by comparison with restriction fragments of known length subjected to electrophoresis on the same gel.

The ability to determine the length of restriction fragments makes it possible to locate the positions of restriction sites relative to one another on a DNA molecule (e.g., a newly cloned DNA fragment). A diagram showing the positions of restriction sites on a DNA molecule is called a *restriction map,* and the process of determining these positions is called *restriction-site mapping.* In this simple case, three fragment samples are digested: one with enzyme I, one with enzyme II, and one with both enzymes.

When a DNA fragment contains *multiple* copies of the recognition site for one or more restriction enzymes, the mapping procedure is more complicated. In this case, the sites for each enzyme must be mapped before the sites for different enzymes can be mapped relative to one another. The first step in this procedure is radiolabeling the 52 ends of both strands of the fragment with [γ-^{32}P]ATP and polynucleotide kinase.The doubly end-labeled fragment is treated with a restriction enzyme that cuts the fragment just once, and the resulting singly end-labeled fragments are separated. These fragments then are partially digested with the enzyme whose multiple recognition sites are being mapped.

Multiple restriction sites in a cloned DNA fragment can be mapped by use of these two methods with multiple restriction enzymes. Because each distinct DNA sequence has a characteristic restriction-site map, such maps can be used to align partially overlapping cloned DNA fragments. Also, specific small regions within a large cloned DNA fragment can be prepared by digesting the cloned fragment with various combinations of restriction enzymes; the smaller subfragments then can be isolated by gel electrophoresis.

PULSED-FIELD GEL ELECTROPHORESIS SEPARATES LARGE DNA MOLECULES

The gel electrophoretic techniques described so far can resolve DNA fragments up to ≈20 kb in length. Larger DNAs, ranging from 2×10^4 to 10^7 base pairs [20 kb to 10 megabases (Mb)] in length, can be separated by size with *pulsed-field gel electrophoresis*. This technique depends on the unique behavior of large DNAs in an electric field that is turned on and off (pulsed) at short intervals.

When an electric field is applied to large DNA molecules in a gel, the molecules migrate in the direction of the field and also stretch out lengthwise. If the current then is stopped, the molecules begin to "relax" into random coils. The time required for relaxation is directly proportional to the length of a molecule. The electric field then is reapplied at 90° or 180° to the first direction. Longer molecules relax less than shorter ones during the time the current is turned off. Since the molecules must relax into a random coil before moving off in a new direction, longer molecules start moving in the direction imposed by the new field more slowly than shorter ones. Repeated alternation of the field direction gradually forces large DNA molecules of different size farther and farther apart.

Pulsed-field gel electrophoresis is very important for purifying long DNA molecules up to ≈10^7 base pairs in length. The technique is required for analyzing cellular chromosomes, which range in size from about 5 × 10^5 base pairs (smallest yeast chromosomes) to 2 –†3 × 10^8 base pairs (animal and plant chromosomes). Very large chromosomes must be digested into fragments of 10^7 base pairs or less before they can be analyzed. Such large restriction fragments can be generated with restriction enzymes that cut at rarely occurring 8-bp restriction sites.

PURIFIED DNA MOLECULES CAN BE SEQUENCED RAPIDLY BY TWO METHODS

Virtually all the information required for the growth and development of an organism is encoded in the DNA of its genome. The availability of techniques to produce and separate DNA restriction fragments a few hundred nucleotides long led to development of two procedures for determining the exact nucleotide sequence of stretches of DNA up to ≈500 nucleotides long. These DNA sequencing methods, together with the technology for constructing a library representing the entire genome of an organism, make it possible to determine the exact sequence of the entire DNA of that organism.

The total genomes of many viruses, several bacteria and archaeans, the yeast *S. cerevisiae,* and the roundworm *C. elegans* have already been sequenced. Automation of the techniques for sequencing DNA and computerized storage of the sequence data are facilitating the current effort to determine the sequence of the entire human genome. Within the next decade, if not sooner, researchers also are likely to complete sequencing the entire genomes of the fruit fly *Drosophila melanogaster,* the mouse, and other important experimental organisms. Knowledge of these DNA sequences will undoubtedly revolutionize our understanding of how cells and organisms function.

MAXAM-GILBERT METHOD

In the late 1970s, A. M. Maxam and W. Gilbert devised the first method for sequencing DNA fragments containing up to ≈500 nucleotides. In this method, four samples of an end-labeled DNA restriction fragment are chemically cleaved at different specific nucleotides. The resulting subfragments are separated by gel electrophoresis, and the labeled fragments are detected by autoradiography.The sequence of the original end-labeled restriction fragment can be determined directly from parallel electrophoretograms of the four samples.

SANGER (DIDEOXY) METHOD

A few years later, F. Sanger and his colleagues developed a second method of DNA sequencing, which now is used much more frequently than the Maxam-Gilbert method. The Sanger method is also called *dideoxy sequencing* because it involves use of 22,32 -dideoxynucleoside triphosphates (ddNTPs), which lack a 32 -hydroxyl group. In this method, the single-stranded DNA to be sequenced serves as the template strand for in vitro DNA synthesis; a synthetic 52 -end-labeled oligodeoxynucleotide is used as the primer.

Four separate polymerization reactions are performed, each with a low concentration of one of the four ddNTPs in addition to higher concentrations of the normal deoxynucleoside triphosphates (dNTPs). In each reaction, the ddNTP is randomly incorporated at the positions of the corresponding dNTP; such addition of a ddNTP terminates polymerization because the absence of a 32 hydroxyl prevents addition of the next nucleotide. The mixture of terminated fragments from each of the four reactions is subjected to gel electrophoresis in parallel; the separated fragments then are detected by autoradiography. The sequence of the original DNA template strand can be read directly from the resulting autoradiogram. Once the sequence for a particular cloned DNA fragment is determined, primers for overlapping fragments can be chemically synthesized based on that sequence. The sequence of a long continuous stretch of DNA thus can be determined by individually sequencing the overlapping cloned DNA fragments that compose it.

BIOINFORMATICS

Vast amounts of DNA sequence have already been determined, and the pace at which new sequences are characterized is continuously accelerating. Computers are necessary to store and distribute this enormous volume of data. *Bioinformatics* is the rapidly developing area of computer science devoted to collecting, organizing, and analyzing DNA and protein sequences. The principal data banks where such sequences are stored are the GenBank at the National Institutes of Health, Bethesda, Maryland, and the EMBL Sequence Data Base at the European Molecular Biology Laboratory in Heidelberg. These databases continuously exchange newly reported sequences and make them available to molecular cell biologists throughout the world on the Internet. Newly derived sequences can be compared with previously determined sequences to search for similarities, called *homologous sequences.* Protein-coding regions can be translated into amino acid sequences, which also can be compared. Because of degeneracy in the genetic code, related proteins often exhibit more homology than the genes encoding them. As we saw in the previous section, databases of partial cDNA sequences (EST databases) are particularly useful in designing probes for screening libraries.

STORED SEQUENCES SUGGEST FUNCTIONS OF NEWLY IDENTIFIED GENES AND PROTEINS

As discussed proteins with similar functions often contain homologous amino acid sequences that correspond to important functional domains in the threedimensional structure of the proteins. Discovery that a protein encoded by a newly cloned gene exhibits such homologies with proteins of known function can provide revealing insights into the function of the cloned gene.

Mutations in *NF1* are associated with the inherited disease neurofibromatosis 1, in which multiple tumors develop in the peripheral nervous system, causing large protuberances in the skin (the "elephant-man" syndrome). Before the *NF1* gene was cloned, there was little understanding of the molecular basis of the disease. After a cDNA clone of *NF1* was isolated and sequenced, the deduced sequence of the NF1 protein was checked against all other protein sequences in GenBank. A region of NF1 protein was discovered to have homology to a yeast protein called Ira. Previous studies in yeast had shown that Ira is a GAP-type protein. These proteins regulate the function of a second type of protein called Ras. This finding was significant because the human *RAS* gene was known to be mutated in many human tumors. As we examine GAP and Ras proteins normally function to control cell replication and differentiation in response to signals from neighboring cells. Because of its sequence homology with GAP, researchers hypothesized that NF1 also would regulate Ras. The NF1 protein subsequently was expressed from the cloned gene by methods described later and, indeed, was found to regulate Ras activity. These findings suggest that in individuals with neurofibromatosis, who have a defective *NF1* gene, a mutant NF1 protein is expressed in cells of the peripheral nervous system; the resulting abnormal interaction between the mutant NF1 and Ras leads to inappropriate cell division and tumor formation.

Another example of this approach involves the *BRCA-1* gene. Women who inherit a mutant form of this gene have a high probability of developing breast cancer before age 50. The *BRCA-1* gene was isolated and a cDNA of the *BRCA-1* mRNA was cloned and sequenced, revealing the amino acid sequence of the BRCA-1 protein. Sophisticated methods of sequence comparison revealed that BRCA-1 protein is distantly, but significantly related to the *S. cerevisiae* Rad9 protein, a cell-cycle checkpoint protein, which functions to control cell division in yeast. Based on this clue, several laboratories have initiated experiments to determine if the BRCA-1 protein functions similarly to the *S. cerevisiae* Rad9 protein.

These examples illustrate how insight into the molecular basis of inherited human diseases can be gained by identifying and cloning the associated mutant gene and then comparing the sequence of the encoded protein with the sequences of other proteins stored in data banks. This general approach for revealing the molecular function of various proteins simply by sequencing the DNA that encodes them will undoubtedly increase as the sequences of more proteins with known functions are determined and as new computer methods are devised for identifying potentially significant relationships between sequences.

COMPARATIVE ANALYSIS OF GENOMES REVEALS MUCH ABOUT AN ORGANISM'S BIOLOGY

The availability of the complete genome sequences of several organisms now permits the comparative analysis of genomes, a branch of bioinformatics known as genomics. With this approach, much can be learned about the biology of organisms that are difficult to culture, even when few if any of their proteins have been isolated and studied directly. For example, Most proteins from three of these organisms – *Mycoplasma genitalium, Methanococcus jannaschii,* and *Haemophilus influenzae* – have not been studied directly. Rather their functions are inferred from their sequences by comparison with previously studied proteins of model organisms like *E. coli* and *S. cerevisiae.*

M. genitalium has the smallest known genome of any cell. Most of the proteins encoded by these long ORFs have been assigned postulated functions based on their sequence homology with proteins of known functions from *E. coli* and other better-studied bacteria. *M. genitalium,* which lives as a parasite within epithelial cells lining primate urogenital and respiratory tracts, is extremely difficult to culture separately from its host eukaryotic cells. Nonetheless, a great deal can be inferred about *M. genitalium* from an analysis of its genome sequence.

For example, the *M. genitalium* genome does not encode many of the enzymes of intermediary metabolism found in free-living organisms (e.g., *E. coli*) that can synthesize all their constituents from simple molecules such as acetate and ammonia plus inorganic ions. The lack of these enzymes suggests that *M. genitalium* produces its macromolecules primarily from preformed precursors that are transported into its cytoplasm from the cytoplasm of its eukaryotic host cell. Although the *M. genitalium* genome lacks recognizable genes encoding the enzymes of the tricarboxylic acid (TCA) cycle and the cytochromes required for oxidative phosphorylation, it does encode the enzymes required for glycolysis of glucose to lactate and acetate. Thus this organism most likely produces ATP principally from the comparatively simple and low-yield glycolytic pathway. Remarkably, the *M. genitalium* genome almost completely lacks the class of proteins that regulate transcription in other bacteria. This implies that the environment within its eukaryotic host cell is so constant that only minimal regulation of gene expression is required for cell growth and division. Alternatively, *M. genitalium* may regulate the expression of its genes by mechanisms that differ substantially from other prokaryotes.

The *M. jannaschii* genome is the first genome from an archaean to be completely sequenced. A strict anaerobe, *M. jannaschii* must live in an oxygen-free environment and derives chemical energy by the reduction of CO_2 with H_2 to produce methane. Genes encoding all the required enzymes for this reductive pathway have been identified in the *M. jannaschii* genome. Comparison of the sequences of *M. jannaschii* proteins with proteins performing equivalent functions from bacteria and eukaryotes has provided insights about the evolution of these three major classes of life on earth, as discussed below.

H. influenzae is one of the first important human pathogens whose genome has been sequenced. It can cause life-threatening bacterial meningitis, an inflammation of the meninges tissue that covers the central nervous system. As a pathogen that normally lives in close association with eukaryotic host cells, *H. influenzae* lacks many of the enzymes of intermediary metabolism present in *E. coli*. For this reason, *H. influenzae*, like *M. genitalium*, cannot grow on minimal medium. An understanding of the complete set of *H. influenzae* genes may provide clues for designing more effective therapies against *H. influenzae* infection. Because of the potential of this approach, scientists are determining the complete genome sequences of a number of pathogenic bacteria.

* An ORF is a DNA sequence that can be divided into triplet codons without any intervening stop codons. Although some polypeptides are shorter than 100 amino acids, these are difficult to predict from DNA sequence alone because *short* ORFs occur randomly in a long DNA sequence. *Long* ORFs encoding 100 or more amino acids are unlikely to occur randomly and very likely encode an expressed polypeptide.

Homologous Proteins Involved in Genetic Information Processing Are Widely Distributed

As noted there are three cell lineages: two prokaryotic lineages, the bacteria and archaea, and one eukaryotic lineage. All cells, regardless of their lineage, process genetic information in three stages: replication, transcription, and translation. Of the three processing systems, translation appears to be the most universal in its overall design in that the majority of proteins associated with ribosomes have clear homologs in bacteria, archaea, and eukaryotes. However, eukaryotic ribosomes, which are larger than prokaryotic ribosomes, contain some proteins not found in either bacteria or archaea. A small number of bacteria-specific ribosomal proteins also occur. In addition, the archaean *M. jannaschii* has homologs of a number of eukaryotic ribosomal proteins that bacteria lack. The *M. jannaschii* translation-initiation and elonga-tion factors and aminoacyl-tRNA synthetases also are more similar to their eukaryotic counterparts. The characteristics of the signal peptides of secreted proteins are also similar in archaea, bacteria, and eukaryotes, suggesting that the basic mechanisms of membrane targeting and translocation are similar among all three cell lineages.

The systems for replicating and transcribing DNA in archaea and eukaryotes are more similar to each other than to the bacterial systems. For example, bacteria lack histones, but the *M. jannaschii* genome encodes five histones homologous to eukaryotic H1, H2A, H2B, H3, and H4. This finding suggests that histones play similar functions in genome packaging and regulation of transcription in both archaea and eukaryotes. Likewise, the genomes of archaea, but not of bacteria, contain genes encoding proteins homologous to the transcription factors that help initiate transcription of nuclear genes in eukaryotic cells. Using methods described later in this chapter, these archaeal proteins have been produced and found to have functions similar to their eukaryotic counterparts. In contrast, the proteins that transcribe the small DNA genomes found in chloroplasts and mitochondria are homologs of the transcription proteins found in bacteria. Mitochondrial and chloroplast ribosomes also resemble bacterial ribosomes more than the cytoplasmic ribosomes of eukaryotes. These new results from genomics imply that the nuclei of modern eukaryotic cells and modern archaea evolved from a common ancestor, whereas mitochondria and chloroplasts evolved from bacteria that formed a symbiotic relationship with an ancient ancestor of the modern eukaryotic cell.

In contrast to histones and transcription proteins, many *M. jannaschii* proteins involved in cell structure are homologous to bacterial structural proteins and are unrelated to eukaryotic proteins. This pattern reflects the similar structural organization of the two classes of prokaryotic cells and their structural differences from eukaryotic cells.

MANY YEAST GENES FUNCTION IN INTRACELLULAR PROTEIN TARGETING AND SECRETION

The genome from the yeast *S. cerevisiae* was the first eukaryotic genome to be sequenced in its entirety. So far, about 65 percent of *S. cerevisiae* genes have known or predicted functions. As shown many basic cell processes are carried out by roughly the same number of proteins in *E. coli* and yeast. The greatest difference occurs in the number of proteins involved in intracellular protein targeting and secretion, with yeast having about 10 times as many proteins devoted to these processes as *E. coli*. This difference reflects the complex intracellular membranes and organelles of eukaryotic cells.

To determine the function of the remaining 35 percent of yeast genes, whose function is not yet known or postulated, scientists have organized a worldwide systematic effort to inactivate each of these genes using methods. Many of the encoded proteins will probably have equivalents in other eukaryotic cells, including human cells. Indeed, about 45 percent of the human genes thus far sequenced have similarity to yeast genes. The recently completed sequence of the genome from *C. elegans*, a small multicellular organism, suggests that many of the human genes without yeast homologs encode proteins required for multicellular animal life.

THE *C. ELEGANS* GENOME ENCODES NUMEROUS PROTEINS SPECIFIC TO MULTICELLULAR ORGANISMS

The sequence of the roundworm *C. elegans* genome, which was virtually completed in 1998, represents the first sequenced genome of a multicellular organism. The 97-Mb genome contains ≈19,100 protein-coding genes, about three times the number of yeast genes and one-fifth to one-third the estimated number of human genes. Approximately 40 percent of the predicted *C. elegans* proteins are homologous to genes of known function from other organisms. By now, ≈5000 human genes have been sequenced, mostly as cDNAs. About 75 percent of the sequenced human genes are homologous to *C. elegans* genes and are expected to have similar functions. This extensive homology makes *C. elegans* an excellent model organism for discovering the functions of many human proteins.

Comparison of the genomes of *S. cerevisiae*, a single-celled eukaryote, and *C. elegans*, a simple multicellular eukaryote, reveals that they encode a comparable number of homologous proteins to carry out the basic eukaryotic cellular functions. However, although *C. elegans* is composed of only ≈1000 cells, its genome also encodes a multitude of proteins that participate in several functions critical to the development and functioning of multicellular organisms. The structures and functions of such proteins, many of which have no homologs in single-celled organisms, are discussed in detail in subsequent chapters.

ANALYZING SPECIFIC NUCLEIC ACIDS IN COMPLEX MIXTURES

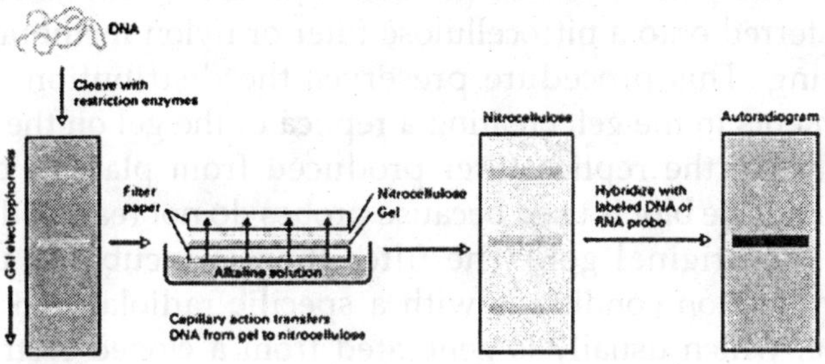

Fig. The Southern blot technique for detecting the presence of specific DNA sequences following gel electrophoresis of a complex mixture of restriction fragments.

Once a specific DNA sequence has been isolated by cloning, the cloned DNA can be used as a probe to detect the presence and the amounts of complementary nucleic acids in complex mixtures such as total cellular DNA or RNA. These procedures depend on the exquisite specificity of nucleic acid hybridization. Related methods are used to locate DNA regions encoding specific mRNAs and transcription start sites.

SOUTHERN BLOTTING DETECTS SPECIFIC DNA FRAGMENTS

The technique of Southern blotting, named after its originator Edwin Southern, can identify specific restriction fragments in a complex mixture of restriction fragments. The DNA to be analyzed, such as the total DNA of an organism, is digested to completion with a restriction enzyme. For an organism with a complex genome, this digestion may generate millions of specific restriction fragments. The complex mixture of fragments is subjected to gel electrophoresis to separate the fragments according to size. However, many different fragments are of exactly the same length, and these do not separate from one another.

Even though all the fragments are not resolved by gel electrophoresis, an individual fragment that is complementary to a specific DNA clone can be detected. The restriction fragments present in the gel are denatured with alkali and transferred onto a nitrocellulose filter or nylon membrane by blotting. This procedure preserves the distribution of the fragments in the gel, creating a replica of the gel on the filter, much like the replica filter produced from plaques of a λ library. (The blot is used because probes do not readily diffuse into the original gel.) The filter then is incubated under hybridization conditions with a specific radiolabeled DNA probe, which usually is generated from a cloned restriction fragment. The DNA restriction fragment that is complementary to the probe hybridizes, and its location on the filter can be revealed by autoradiography.

Southern blotting permits a comparison between the restriction map of DNA isolated directly from an organism and the restriction map of cloned DNA. This comparison is necessary to be certain that no rearrangements have occurred during the cloning procedure such as might happen if two restriction fragments that do not normally lie next to each other were inadvertently ligated together before ligation into a cloning vector. Southern blotting also is used to map restriction sites in genomic DNA next to the sequence of a cloned DNA fragment. This provides a rapid method of comparing the restriction maps of different individual organisms in the region surrounding a cloned fragment. Deletion and insertion mutations are readily detected, as well as sequence differences in specific restriction sites.

NORTHERN BLOTTING DETECTS SPECIFIC RNAS

Northern blotting, humorously named because it is patterned after Southern blotting, is used to detect a particular

RNA in a mixture of RNAs. An RNA sample, often the total cellular RNA, is denatured by treatment with an agent (e.g., formaldehyde) that prevents hydrogen bonding between base pairs, ensuring that all the RNA molecules have an unfolded, linear conformation. The individual RNAs then are separated according to size by gel electrophoresis and transferred to a nitrocellulose filter to which the extended denatured RNAs adhere. The filter then is exposed to a labeled DNA probe and subjected to autoradiography. Because the amount of a specific RNA in a sample can be estimated from a Northern blot, the procedure is widely used to compare the amounts of a particular mRNA in cells under different conditions.

SPECIFIC RNAS CAN BE QUANTITATED AND MAPPED ON DNA BY NUCLEASE PROTECTION

Another important method for detecting and quantitating specific RNA molecules employs endonucleases that digest single-stranded but not double-stranded nucleic acids. The method was originally designed using endonuclease S1, an enzyme from the mold *Aspergillus oryzae* that digests singlestranded RNA and DNA but not double-stranded molecules. A labeled DNA strand, or probe, complementary to an RNA of interest is prepared from a cloned DNA. A source of RNA, such as the total polyadenylated RNA isolated from a particular tissue or type of cultured cell, is incubated with a high concentration of the labeled DNA probe under conditions in which all the RNA complementary to the probe hybridizes to it. The preparation then is treated with endonuclease S1, which digests all the unhybridized RNA and probe molecules, leaving only the double-stranded region in the RNA-DNA hybrids, which is protected from nuclease digestion. Treatment of the digested preparation with alcohol precipitates the probe –†target RNA hybrid, which then is subjected to gel electrophoresis followed by autoradiography or phosphorimager analysis of the gel to detect the protected probe. The amount of radioactivity in the resulting band is a measure of the amount of RNA complementary to the probe in the initial sample of RNA. Nuclease protection also can be performed with a complementary labeled RNA and ribonuclease A, a single-strand-specific pancreatic ribonuclease.

The DNA region that encodes a particular RNA can be mapped with the nuclease-protection technique by use of restriction-fragment probes in which one end is complementary to only a portion of the RNA of interest. In this case, the RNA-DNA hybrid protected from S1 digestion is shorter than the RNA being probed; its length corresponds to that of the DNA region extending from one end of the coding region to a restriction site within it. Comparison of the protected doubled-stranded fragments obtained with two or more such "partial" probes can map the RNA sequence relative to restriction sites in the complementary DNA.

TRANSCRIPTION START SITES CAN BE MAPPED BY S1 PROTECTION AND PRIMER EXTENSION

As discussed some of the DNA regulatory elements that control transcription of genes into mRNA are located near the transcription *start site.* Mapping the start site for synthesis of a particular mRNA often helps in identifying the DNA regulatory sequences that control its transcription. Two methods are used to map the 52 end of a particular mRNA on a complementary DNA: *S1 protection* and *primer extension.* The first step in both methods is to identify the general region of a DNA that includes the start site of interest ; this can be done by Northern blot analysis or nuclease protection using various cloned restriction fragments as probes.

In the S1-protection method, the identified DNA region is treated with appropriate restriction enzymes to produce a single-stranded DNA fragment that will hybridize with the 52 portion of the mRNA. This fragment is radiolabeled at the 52 end, hybridized with the mRNA, and then trimmed with S1 endonuclease. From the length of the labeled probe segment protected from digestion, the position of the start site in the original DNA can be located.

The primer-extension method uses a synthetic oligonucleotide that is complementary to an approximately 20-

nucleotide stretch of the mRNA located 50 – 200 nucleotides from its 52 end. This synthetic oligonucleotide is end-labeled at the 52 end and then used to prime DNA synthesis by reverse transcriptase with the mRNA as the template. The position of the start site can be mapped from the length of the resulting extension product.

PRODUCING HIGH LEVELS OF PROTEINS FROM CLONED CDNAS

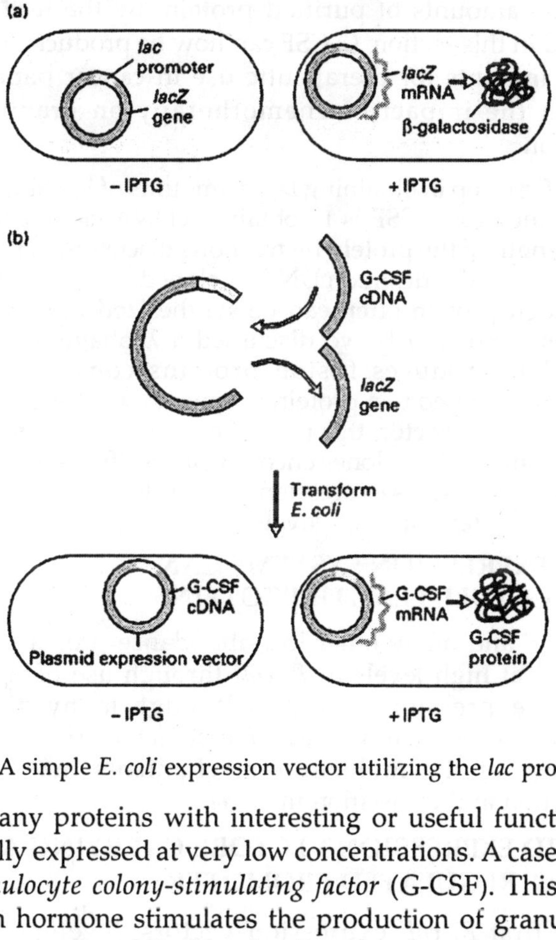

Fig. A simple *E. coli* expression vector utilizing the *lac* promoter.

Many proteins with interesting or useful functions are normally expressed at very low concentrations. A case in point is *granulocyte colony-stimulating factor* (G-CSF). This human protein hormone stimulates the production of granulocytes, the phagocytic white blood cells critical to defence against bacterial infections. Both granulocytes and the cells that produce G-CSF are very sensitive to chemotherapeutic agents used in the treatment of cancer. As a result, one of the most serious side effects of chemotherapy is a fall in the concentration of granulocytes, making patients prone to life-threatening infections. G-CSF is normally made in very low concentrations in the bone marrow, where granulocytes differentiate and by some cultured human cell lines. Isolation of G-CSF from these sources is a tedious process yielding minuscule amounts of purified protein. By the techniques described in this section, G-CSF can now be produced in large enough amounts for therapeutic use in cancer patients to diminish the impact of chemotherapy on granulocyte production.

The first step in obtaining large amounts of lowabundance proteins such as G-CSF is to obtain a cDNA clone encoding the full length of the protein by methods discussed in previous sections. Once the desired cDNA is cloned, large amounts of the encoded protein often can be synthesized in engineered *E. coli* cells. Previously we discussed a λ phage expression vector that produces fusion proteins consisting of β-galactosidase joined to a protein fragment encoded by a DNA inserted into the vector; this type of expression system can be used to identify DNA clones encoding a specific protein. Here, in contrast, we discuss expression vectors designed to produce full-length proteins at high levels.

E. COLI EXPRESSION SYSTEMS CAN PRODUCE FULL-LENGTH PROTEINS

G-CSF and many other low-abundance proteins can be expressed at high levels in *E. coli* through use of specially designed expression vectors. Although many different expression vectors have been constructed, they all take advantage of the molecular mechanisms that control transcription and translation in *E. coli.*

PLASMID EXPRESSION VECTORS CARRYING A STRONG, REGULATED PROMOTER

The first *E. coli* expression vectors developed were assembled by ligation of a basic plasmid vector, containing a replication origin (ORI) and selectable antibiotic-resistance gene, to a DNA sequence that functions as a strong, regulated promoter. A promoter is a DNA sequence where RNA polymerase initiates transcription. At a strong, regulated promoter, transcription is initiated many times per minute under specific environmental conditions. For example, one expression vector contains a cloned fragment of the *E. coli* chromosome that includes the *lac* promoter and the *lacZ* gene encoding β-galactosidase. Transcription from the *lac* promoter occurs only when lactose, or a lactose analog such as isopropylthiogalactoside (IPTG), is added to the culture medium. IPTG generally is used because it cannot be metabolized, and therefore its concentration does not change as the cells grow. After addition of IPTG, the *lacZ* gene is transcribed into mRNA, which then is translated to yield many copies of the β-galactosidase protein.

To modify this plasmid for production of G-CSF, the *lacZ* gene is replaced with a cDNA encoding G-CSF using restriction enzymes and DNA ligase. In this process, the *lac* promoter, which is required for efficient transcription, must be maintained just before the start site of the inserted cDNA. In *E. coli* cells transformed by the resulting plasmid, transcription of the G-CSF cDNA and expression of G-CSF protein occurs in the presence of IPTG.

PLASMID EXPRESSION VECTORS CARRYING THE T7 LATE PROMOTER

A more complicated expression system, involving two levels of amplification, can produce larger amounts of a desired protein than the system just described. This second-generation system depends on the regulated expression of *T7 RNA polymerase,* an extremely active enzyme that is encoded in the DNA of bacteriophage T7. The T7 RNA polymerase transcribes DNA beginning within a specific 23-bp promoter sequence called the *T7 late promoter.* Copies of the T7 late promoter are located at several sites on the T7 genome, but none is present in *E. coli* chromosomal DNA. As a result, in

T7-infected cells, T7 RNA polymerase catalyzes transcription of viral genes but not of *E. coli* genes.

In this expression system, recombinant *E. coli* cells first are engineered to carry the gene encoding T7 RNA polymerase next to the *lac* promoter. These cells then are transformed with plasmid vectors that carry a copy of the T7 late promoter and, adjacent to it, the cDNA encoding the desired protein. When IPTG is added to the culture medium containing these transformed, recombinant *E. coli* cells, T7 RNA polymerase is expressed by transcription from the *lac* promoter. The polymerase then binds to the T7 late promoter on the plasmid expression vectors, catalyzing transcription of the inserted cDNA at a high rate. Since each *E. coli* cell contains many copies of the expression vector, prodigious amounts of mRNA corresponding to the cloned cDNA can be produced in this system. Typically 10 – 70 percent of the total protein synthesized by these cells after addition of IPTG is the protein of interest. Because of the high yields possible with the T7 two-step expression system, it is often used for producing proteins in *E. coli*.

EUKARYOTIC EXPRESSION SYSTEMS CAN PRODUCE PROTEINS WITH POST-TRANSLATIONAL MODIFICATIONS

Most of the enzymes used in recombinant DNA technology (e.g., restriction enzymes, DNA polymerases, DNA ligases, polynucleotide kinase, and reverse transcriptase) are now produced commercially in *E. coli* expression systems. Large quantities of many eukaryotic proteins of interest, such as G-CSF, also can be produced in these systems. However, some eukaryotic proteins cannot be produced in active form in *E. coli* cells. These include proteins that are extensively modified during or following their synthesis, such as glycoproteins to which carbohydrate groups are added. *E. coli* lacks the enzymes that catalyze many of the post-translational modifications found on eukaryotic proteins.

To overcome this limitation of *E. coli* expression systems, researchers have developed *eukaryotic expression vectors* that permit addition of appropriate post-translational modifications to expressed proteins. Such vectors can be used in various types of eukaryotic cells to direct abundant synthesis of eukaryotic proteins from cloned genes.

CLONED CDNAS CAN BE TRANSLATED IN VITRO TO YIELD LABELED PROTEINS

In the *E. coli* and eukaryotic expression systems just described, proteins are synthesized in vivo within living cells. In vitro expression of proteins encoded by cloned cDNA also is possible with the T7 late promoter expression vector diagrammed. In this system, the mRNA encoded by the cloned cDNA is synthesized in vitro using the purified expression vector and purified T7 RNA polymerase. The mRNA then is translated in vitro to yield the desired protein. Although an in vitro expression system yields less protein than an in vivo system, it allows the desired protein to be radioactively labeled by addition of labeled amino acids to the in vitro translation reaction. In addition to the T7 system, plasmid vectors containing late promoters and the corresponding RNA polymerases from related bacteriophages (e.g., T3 and SP6) also can be used for in vitro production of proteins from cloned DNA.

Cell extracts are used to translate mRNA synthesized in vitro into protein. These extracts, which usually are prepared from rabbit reticulocytes or wheat germ, are rich in the components required for translation: ribosome subunits; tRNAs; aminoacyl-tRNA synthetases; and initiation, elongation, and termination factors. Extracts first are treated with low concentrations of micrococcal nuclease to eliminate any endogenous mRNAs. At low levels, this enzyme, which is active only in the presence of Ca^{2+}, digests mRNA but not tRNA or rRNA. After digestion of endogenous mRNAs is complete, the enzyme is inactivated by addition of EGTA, a chelating-agent that binds Ca^{2+} with a much higher affinity than the nuclease does. When an in vitro synthesized mRNA is added to a treated extract, only the corre-sponding protein is produced, since the endogenous mRNAs have been destroyed. If a radiolabeled amino acid such as [^{35}S]methionine is included in the translation reaction, the protein product will be labeled. The labeled protein can be used in binding experiments to test its interaction with other proteins and in various other experimental systems.

POLYMERASE CHAIN REACTION: AN ALTERNATIVE TO CLONING

An alternative to cloning, called the polymerase chain reaction(PCR), can be used to directly amplify rare specific DNA sequences in a complex mixture when the ends of the sequence are known. This method of amplifying rare sequences from a mixture has numerous applications in basic research, human genetics testing, and forensics.

Genomic DNA is digested into large fragments using a restriction enzyme and then is heat-denatured into single strands. Two synthetic oligonucleotides complementary to the 32 ends of the target DNA segment of interest are added in great excess to the denatured DNA, and the temperature is lowered to 50 – 60 °C. The genomic DNA remains denatured, because the complementary strands are at too low a concentration to encounter each other during the period of incubation, but the specific oligonucleotides, which are at a very high concentration, hybridize with their complementary sequences in the genomic DNA. The hybridized oligonucleotides then serve as primers for DNA chain synthesis, which begins upon addition of a supply of deoxynucleotides and a temperature-resistant DNA polymerase such as that from *Thermus aquaticus* (a bacterium that lives in hot springs). This enzyme, called *Taq polymerase,* can extend the primers at temperatures up to 72 °C. When synthesis is complete, the whole mixture is heated further (to 95 °C) to melt the newly formed DNA duplexes. When the temperature is lowered again, another round of synthesis takes place because excess primer is still present. Repeated cycles of synthesis (cooling) and melting (heating) quickly amplify the sequence of interest. At each round, the number of copies of the sequence between the primer sites is doubled; therefore, the desired sequence increases exponentially.

PCR AMPLIFICATION OF MUTANT ALLELES PERMITS THEIR DETECTION IN SMALL SAMPLES

The sensitivity of procedures used in human genetics testing has been vastly increased by use of PCR amplification. For example, the β-globin gene in a small sample of DNA isolated from an individual can be specifically amplified by the PCR to determine if the person is a carrier of the mutant sickle-cell allele. Quantities of amplified DNA sufficient for sequencing can be prepared rapidly; subsequent sequencing reveals if the mutant allele is present in the sample.

The PCR is so effective at amplifying specific DNA sequences that DNA isolated from a single human cell can be analyzed for mutations associated with various genetic diseases. In one reported case, this approach was used to screen in vitro fertilized human embryos prepared from sperm and ova from a couple who both were carriers of the genetic disorder cystic fibrosis. This disease results from mutation in the *CFTR* gene, which is located on chromosome 7. The DNA isolated from a single embryonic cell was subjected to PCR amplification and then analyzed for mutations identified in one of the two copies of chromosome 7 in each parent. In this way embryos that had inherited the wild-type chromosome from at least one parent were identified and then transferred to the mother's uterus. (Removal of a single cell from an in vitro fertilized human embryo has no apparent effect on subsequent development of the embryo after it is implanted in a receptive uterus.) By use of this procedure, carrier couples can be assured of having children that will not be at risk for cystic fibrosis.

Another medical application of the PCR is early detection of infection with HIV, the virus that causes acquired immunodeficiency syndrome (AIDS). The PCR is so sensitive that it can detect HIV at very early stages in the disease (before symptoms appear) when only a few thousand blood cells in a patient are infected with the virus.

DNA SEQUENCES CAN BE AMPLIFIED FOR USE IN CLONING AND AS PROBES

In basic research, the PCR also has numerous applications. For example, this procedure allows the recovery and rapid amplification of the entire DNA region between any two ends whose sequences are known; the amplified DNA fragment then can be ligated into standard cloning vectors. Fragments of ≈2 kb or less can be amplified readily, and recent refinements of the technique allow amplification of regions of >30 kb.

The PCR also provides an alternative approach for preparing probes to screen genomic or cDNA libraries for clones encoding a protein of interest. The amino acid sequence of two peptides isolated from the purified protein are used to design two degenerate oligonucleotide mixtures containing all possible DNA sequences encoding the two peptides. Rather than using these oligonucleotides as probes for direct screening of a cDNA library, as described previously, they are used as primers in a PCR. First, cDNA is synthesized from total cellular mRNA using reverse transcriptase. The cDNA is then used as the template for a PCR performed with the two degenerate oligonucleotide primers. This reaction amplifies the region of the cDNA between the sequences encoding the peptides used to design the degenerate primers. The PCR procedure effectively selects the correct oligonucleotides for priming DNA synthesis from the degenerate oligonucleotide mixtures, because only DNA synthesized from the correct cDNA template will hybridize to oligonucleotides present in both degenerate primer mixtures. For exponential amplification to take place, priming must occur from both ends of a fragment. Even if an oligonucleotide in one of the degenerate mixtures hybridizes to an incorrect cDNA and primes DNA synthesis, the DNA strand that is synthesized will not be amplified, because it will not contain a sequence complementary to one of the oligonucleotides in the second degenerate primer mixture. The cDNA sequence amplified by this procedure contains the unique sequence of the naturally occurring mRNA encoding the region between the two peptides originally sequenced. This unique DNA sequence can then be radioactively labeled and used as a probe for screening a cDNA or genomic library. A probe prepared in this way gives a much stronger and more specific signal than that obtained by direct use of a degenerate probe.

DNA MICROARRAYS: ANALYZING GENOMEWIDE EXPRESSION

Recently invented methods allow researchers to analyse the expression of thousands of genes simultaneously using DNA microarrays. Coupling these methods with the results from genome sequencing projects allows researchers to analyse the complete transcriptional programme of an organism during specific physiological responses or developmental processes. DNA microarrays consist of thousands of individual gene sequences bound to closely spaced regions on the surface of a glass microscope slide. Two methods have been developed for preparing DNA microarrays.

In one method, an ≈1-kb portion of the coding region of each gene analyzed is individually amplified by PCR. A robotic device is used to apply each amplified DNA sample to closely spaced spots on the surface of a glass microscope slide, which then is processed by chemical and heat treatment to attach the DNA sequences to the glass surface and denature them. Typical arrays are 2 × 2 cm and contain ≈6000 spots of DNA.

In an alternative method, multiple DNA oligonucleotides, usually 20 nucleotides in length, are synthesized from an initial nucleotide that is covalently bound to the surface of a glass slide. Tens of thousands of identical oligonucleotides are synthesized in a small square area on the surface of the slide. Several oligonucleotide sequences from a single gene are synthesized in neighboring regions of the slide to analyse expression of that gene. Thousands of genes can be represented on one glass slide. Because the methods for constructing these arrays of synthetic oligonucleotides were adapted from methods for manufacturing microscopic integrated circuits used in computers, these types of oligonucleotide microarrays are often called *DNA chips*.

In one of the first uses of DNA microarray analysis, gene expression was compared between yeast cells growing on glucose as the source of carbon and energy, and cells growing on ethanol. The total mRNA was separately isolated from cells growing under these two conditions using oligo-dT columns The total mRNA from the glucose-grown cells was reverse-transcribed into cDNA using reverse transcriptase, an oligo-dT primer, deoxyribonucleoside triphosphates, and a low concentration of a nucleotide analog labeled with a green fluorescent dye. cDNA from the ethanol-grown cells was prepared similarly, but labeled with a red fluorescent dye. Equal amounts of the two cDNA preparations were then mixed and incubated under hybridization conditions with a DNA microarray containing all 6400 yeast genes, so that each gene on the array hybridized to its complementary cDNA in the preparation.

After washing away unhybridized cDNA, each DNA spot on the microarray was analyzed using a scanning laser microscope with a computer-controlled stage. The intensity of green and red fluorescence at each DNA spot was mea-sured with photomultiplier tubes and stored in computer files under the name of each gene according to its known position on the slide. Spots fluorescing bright red are from genes expressed at a high level in ethanol and a low level in glucose media. Spots fluorescing bright green are from genes expressed at a high level in glucose and a low level in ethanol medium. Spots fluorescing bright yellow are from genes expressed at high levels in both types of media. Genes expressed at low to moderate levels do not yield brightly fluorescing spots in the depiction of the data where all genes are compared. However, the relative level of expression of such genes in cells grown in glucose or ethanol can be determined from accurate measurements of the relative intensities of weak red and green fluorescence signals. Genes that are not transcribed under these growth conditions give no detectable signal.

Hybridization of fluorescently labeled cDNA preparations to DNA microarrays such as this provide a means for analyzing gene expression patterns on a genomic scale. One example is the analysis of changes in gene expression as yeast cells convert from growth on a high-sugar medium to growth on ethanol. This adaptation to different nutrients occurs in the natural environment when yeast growing on decaying fruit switch from rapid growth, when they metabolize sugars to ethanol by anaerobic fermentation, to slower growth, when the sugar is depleted and they metabolize the accumulated ethanol to carbon dioxide and water by aerobic respiration. This situation can be simulated in the laboratory by inoculating yeast into media high in glucose. Growth initially is exponential and rapid until the glucose is exhausted; then the rate of cell doubling is slowed and cells shift from fermentation to respiration. The transcriptional response of each of the 6400 yeast genes during this adaptation to decreasing glucose was analyzed by hybridizing fluorescently labeled cDNAs prepared from cells at several time points after inoculation of the culture to DNA microarrays.

This analysis revealed a marked, coordinated change in the global pattern of yeast gene expression. For example, expression from seventeen genes encoding mitochondrial proteins required for respiration increased with a similar time course as glucose was depleted. Analysis of the yeast genome sequence showed that each of these genes contains in their upstream transcription-control region binding sites for the trimeric transcription factor encoded by genes *HAP2,3,* and *4*. Earlier studies showed that this factor regulates the transcription of some of these genes. This genome-wide expression analysis revealed that these seventeen genes are probably chiefly regulated by the trimeric Hap transcription factor. As the rate of cell growth slowed when glucose was depleted from the media, 112 genes encoding ribosomal proteins were repressed with similar kinetics. The upstream control regions in all of these genes have binding sites for the Rap1 transcription factor, which earlier had been shown to regulate the expression of a handful of ribosomal protein genes. Seven genes exhibited strong induction only at the last time point analyzed, when glucose was completely depleted from the media. Six of these were found to have an upstream activating sequence in common, the CSRE carbon-source response element. Thus, several distinct temporal patterns of expression were recognized. In most cases, common regulatory mechanisms could be inferred for sets of genes with similar expression profiles.

The results of this study showed that as yeast adapt to decreasing glucose levels, expression of 710 genes increases by a factor of two or more, while expression of 1030 genes decreases by a factor of two or more. About 400 of the differentially expressed genes have no known function. This analysis provides the first clue as to their possible function in yeast biology.

Approximately 15,000 human genes have now been identified. Microarrays of these human genes have been prepared and used to characterize their expression in different tissues and organs, and in cancer cells compared to their normal cell counterparts. When the complete set of genes for humans and model experimental organisms such as the mouse and *Drosophila* become available, microarray technology will be used to analyse the global genome-wide gene expression patterns that occur during cellular differentiation and disease.